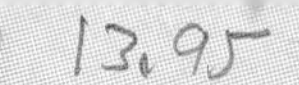
13.95

D0906485

INVISIBLE WORLD

THE MIRACLE OF LIFE

Text: Mercè Parramón
Illustrations: Francisco Arredondo

1 3 5 7 9 8 6 4 2

ISBN 0-7910-2130-0
ISBN 0-7910-2135-1 (pbk.)

Contents

INVISIBLE WORLD

THE MIRACLE OF LIFE

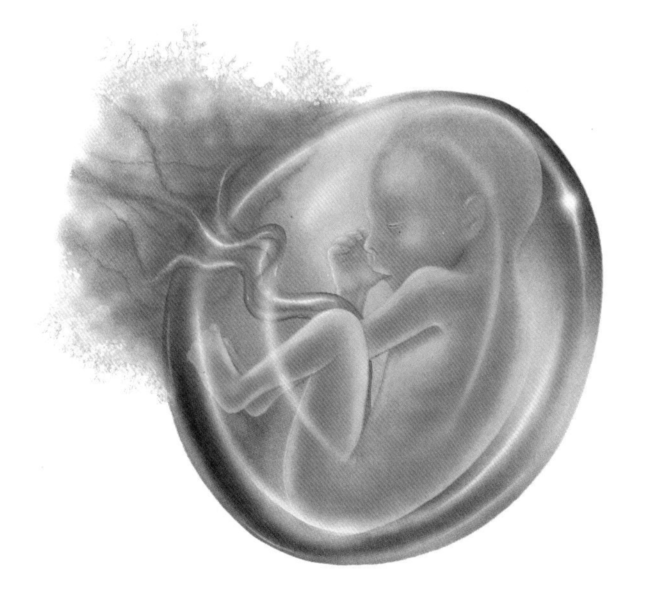

CHELSEA HOUSE PUBLISHERS

New York • Philadelphia

How Life Begins

Reproduction is the process by which living things give life and birth. The first step in reproduction is the joining of two sex cells, one of which comes from the father, and one which comes from the mother.

The male's sex cell is called a spermatozoon, or sperm for short. The female's sex cell is called an ovum, or egg. Reproduction begins when a sperm joins with an ovum to make a single new cell, the zygote. This joining process is called fertilization.

The zygote is smaller than the head of a pin. Yet it contains all the information necessary for producing a new being: the color of the hair, the length of the legs, the number of teeth, the color and shape of the eyes, the shape of the nose, and so on. This information is carried by chromosomes in the cell's nucleus and is passed on to every cell of the new being.

In birds and many egg-laying animals, most of the new being's development takes place outside the mother's body. The young of such animals as elephants, mice, and humans develop inside the mother, until they are ready to be born.

The miracle of life depends upon two tiny cells. The small diagram (below right) shows clearly what is happening in the big picture. Sperm ***(1)*** *struggle to reach a single egg, or ovum* ***(2).*** *Only one of them* ***(3)*** *makes its way into the egg and fertilizes it. The fertilized egg will then divide and develop inside the mother's body.*

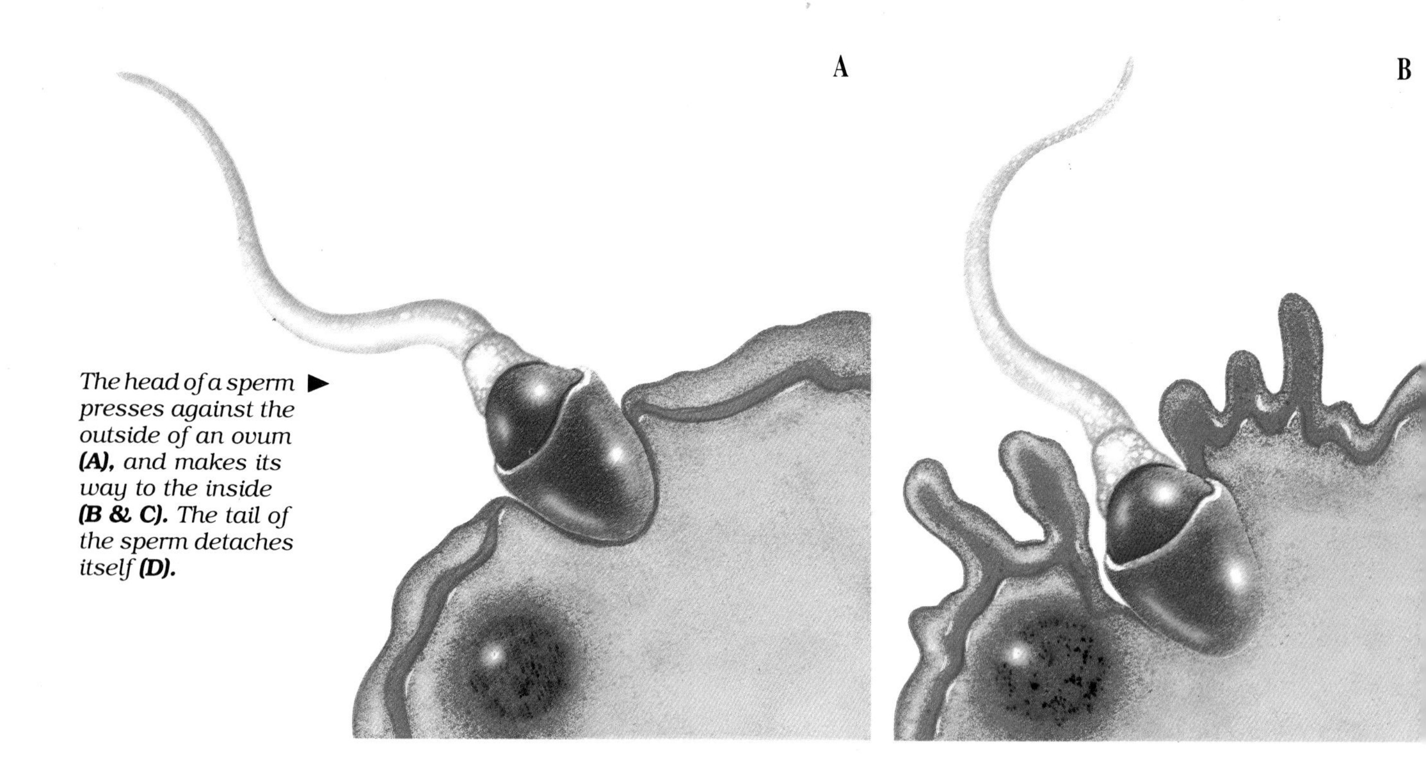

The head of a sperm presses against the outside of an ovum ***(A),*** *and makes its way to the inside* ***(B & C).*** *The tail of the sperm detaches itself* ***(D).***

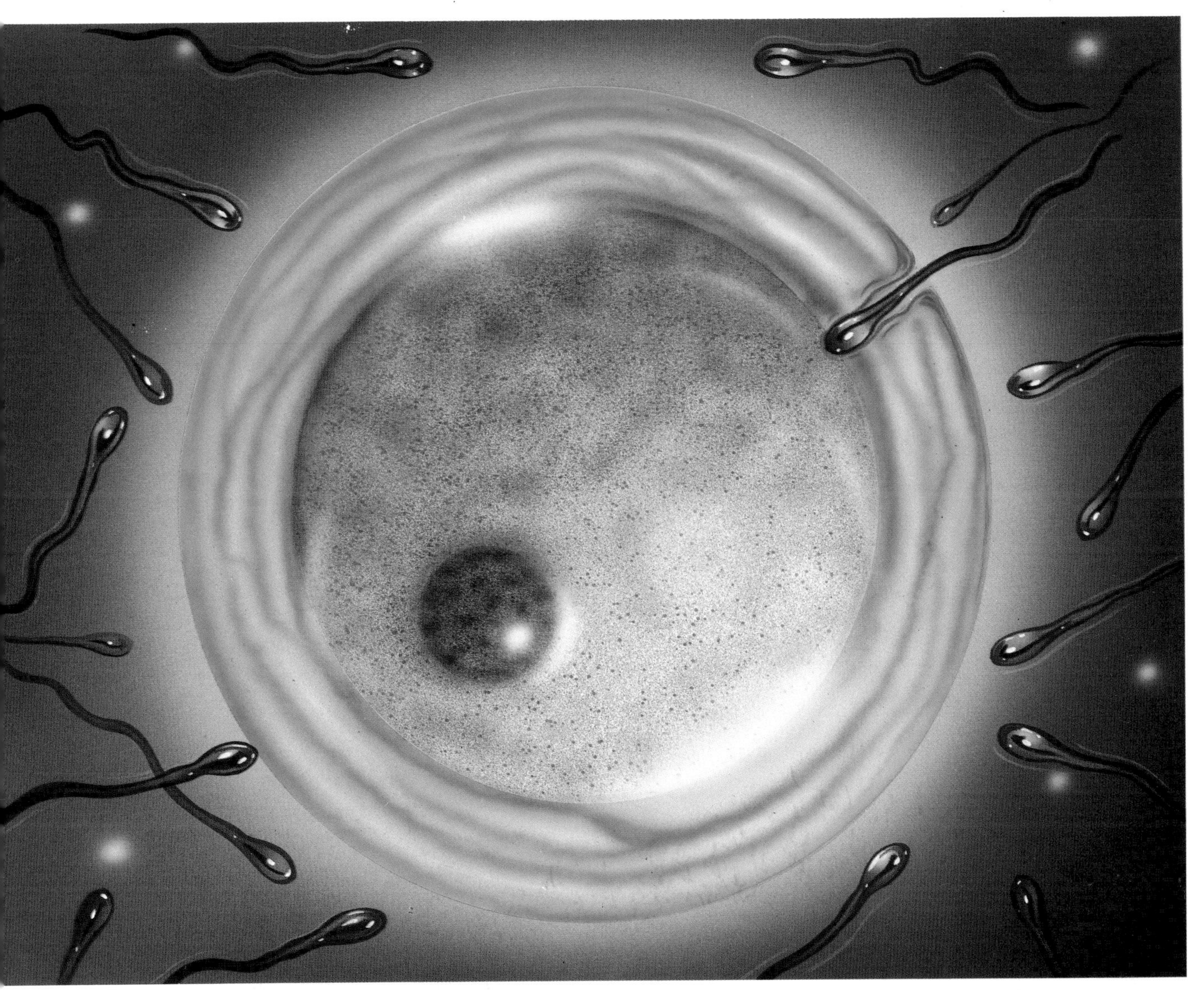

C

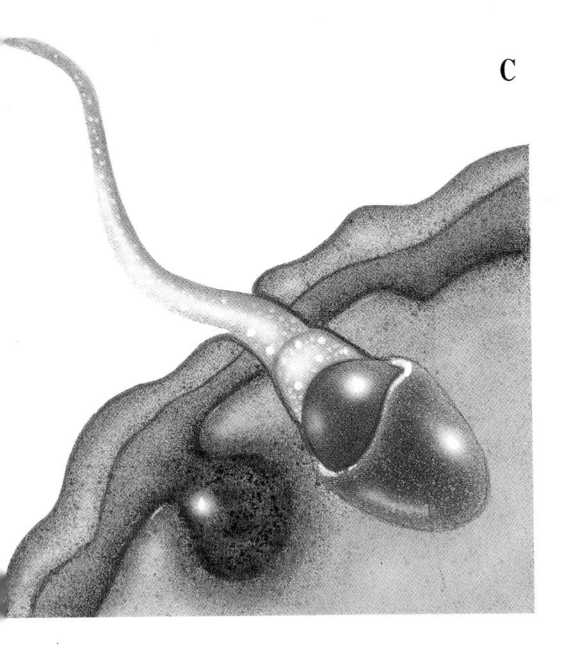

D

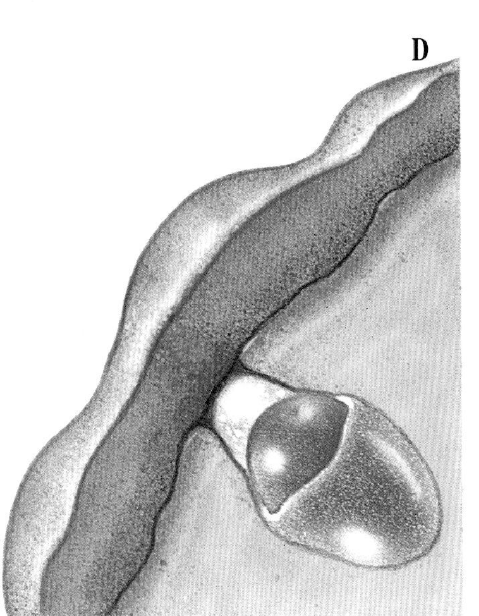

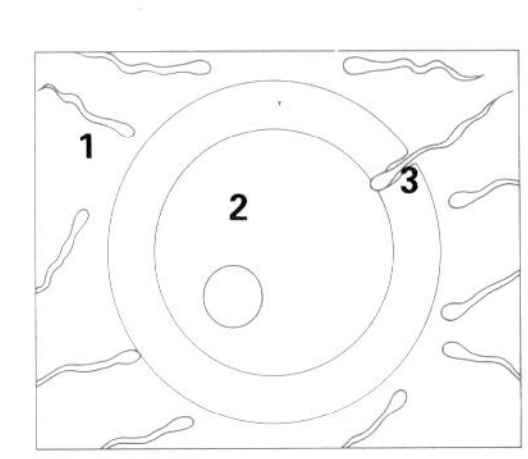

The Female Reproductive System

A woman's reproductive organs (red) are located in her abdomen and are protected by her hip bones.
▼

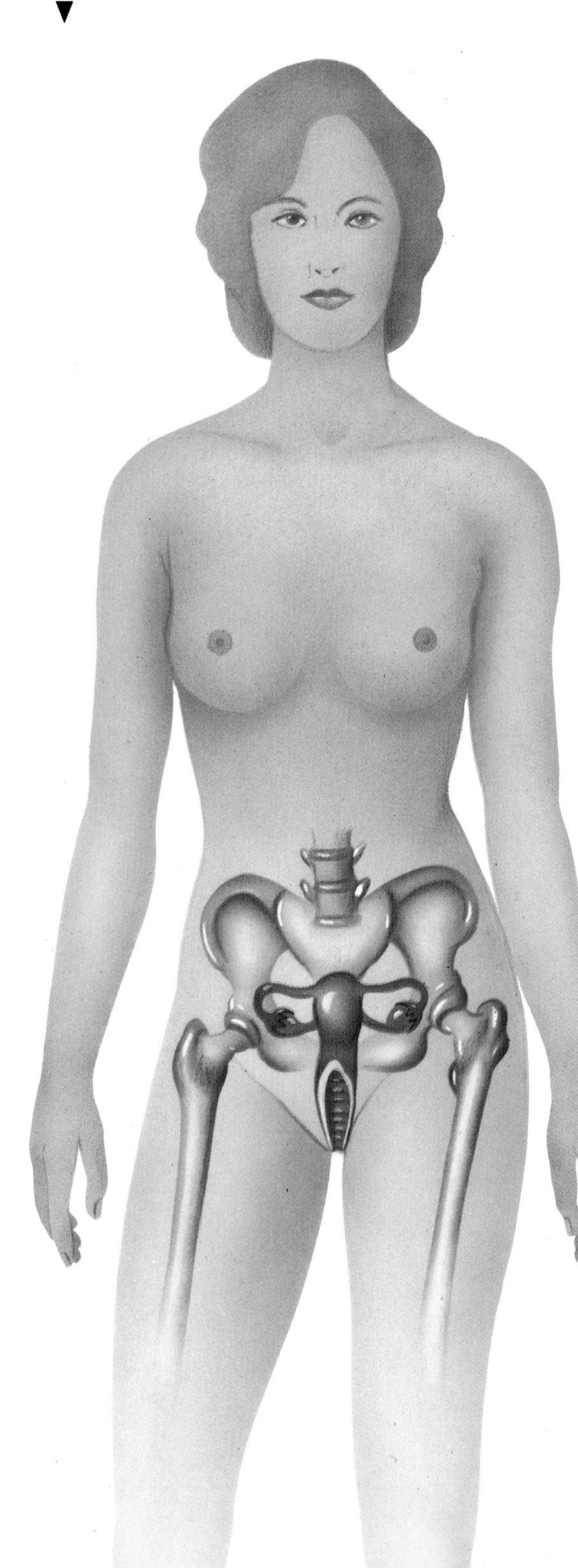

A woman's reproductive system is located in her abdomen, where her baby can grow in safety. Her eggs are formed in her two ovaries, which are situated in the region between the hip bones. An ovary is rather similar to an almond in shape and size.

Inside each ovary are many thousands of egg cells. A woman is born with more than enough eggs to last all her life. Every month, from puberty (which occurs between the ages of about 11 and 14) to menopause (from about 45 to 55), one or more of these cells matures into an ovum. This process is controlled by a hormone called estrogen, which is produced in the ovaries. Estrogen also controls the development of the sex characteristics that make females different from males.

Each month, a mature ovum leaves its ovary and passes into one of the oviducts, or Fallopian tubes, that connect each ovary to the womb, or uterus. The uterus has muscular walls and is roughly the shape and size of a small pear. When an ovum is fertilized, it attaches itself to the wall of the uterus and grows there, taking nourishment from the mother, until it is ready to be born about 9 months later.

The lower end of the uterus leads to the vagina, a canal about 3 to 3.5 inches (7 to 9 centimeters) long that connects the uterus with the outside of the body.

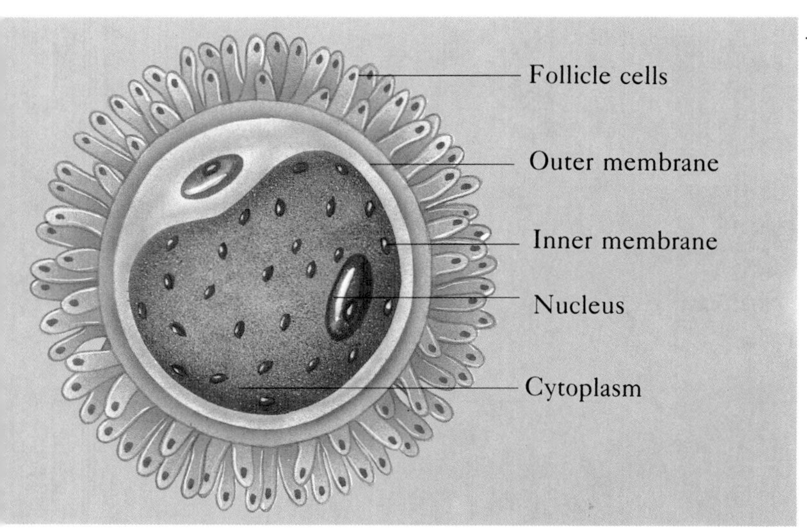

An ovum is surrounded by frond-like follicle cells that help it move. Two tough membranes protect the inside, which is made of a jelly-like substance called cytoplasm. The nucleus is the control center of the cell.

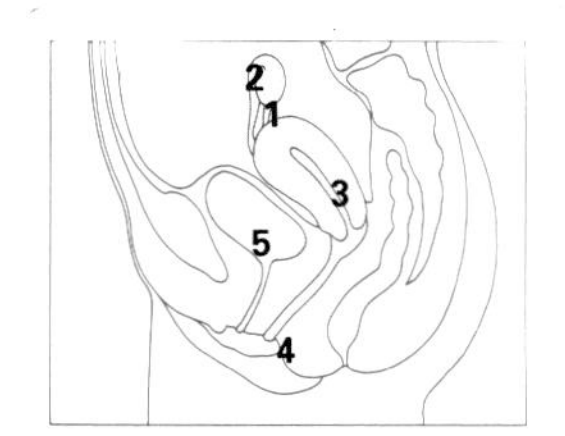

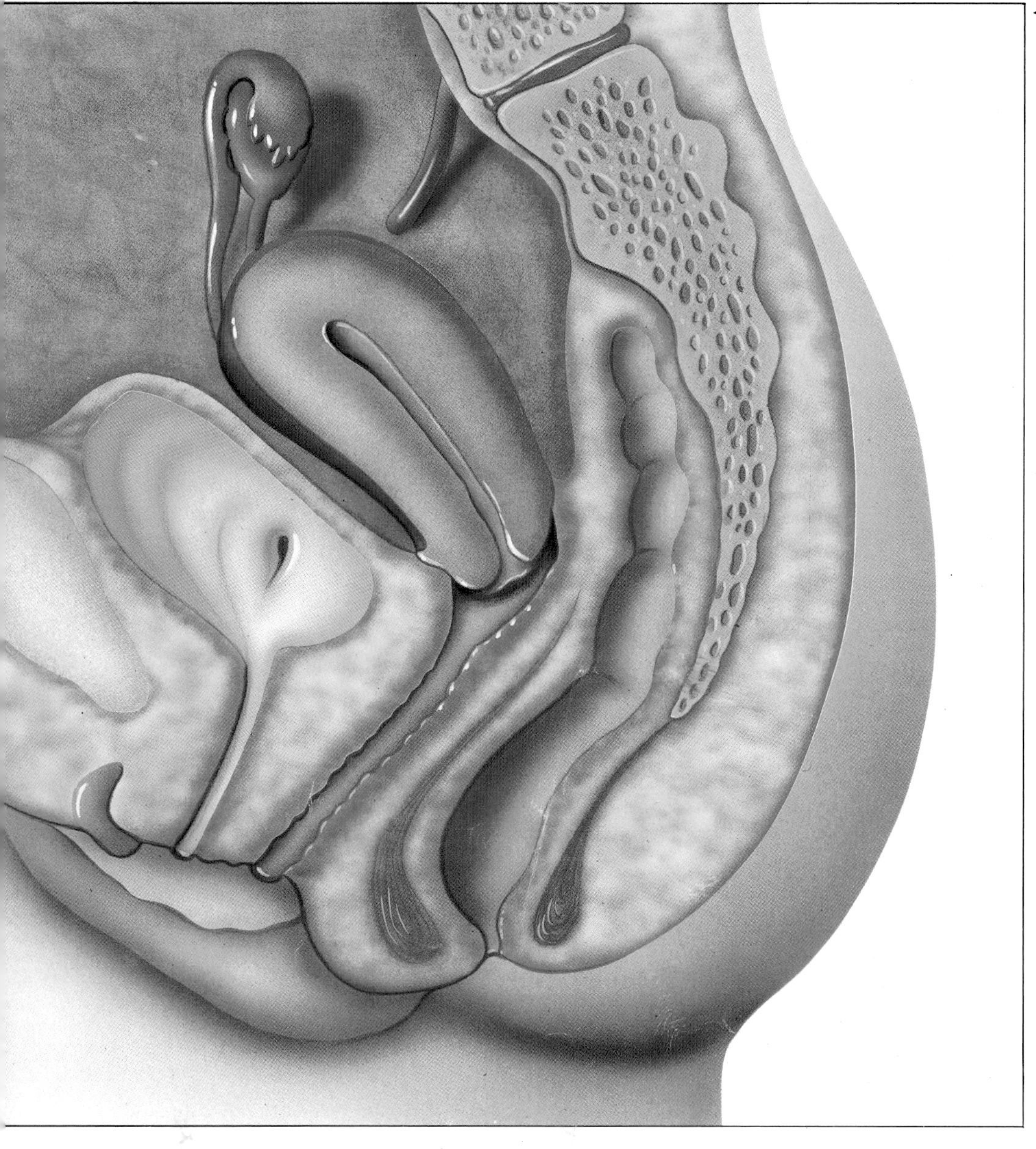

The female reproductive system. Its parts are numbered in the small diagram (above). The almond-shaped ovary ***(1)*** *is partly covered by the fringed, trumpet-shaped end of the Fallopian tube* ***(2),*** *or oviduct, which leads to the uterus* ***(3),*** *where the ovum will grow if it is fertilized. The vagina* ***(4)*** *connects the uterus to the outside of the body. The female reproductive organs lie just behind, and slightly above, the bladder* ***(5).***

The Male Reproductive System

A man's penis and scrotum are outside his body, which keeps his testicles slightly cooler than the rest of his body and is necessary for the production of healthy sperm.
▼

Some of the organs in a man's reproductive system are located outside his body. His organs are designed to produce sperm and place it inside a woman's body.

Sperm is formed in the testicles, which are two glands that lie outside of the body, in a pouch of skin called the scrotum. Each testicle contains a large number of tubes called seminiferous tubules, in which sperm is formed. Between these tubes are cells that produce the hormone testosterone, which controls the male reproductive system and the development of the sex characteristics that make men different from women.

Sperm is stored in the testicles in a coiled tube called the epididymis. From there, sperm makes its way out of the body through a long tube called the vas deferens to the urethra. The sperm is mixed with a liquid called seminal fluid, which is made by the prostate gland and the seminal vesicles, to form semen. There are usually 100 to 200 million sperm in a cubic centimeter of semen.

The urethra is a duct leading to the outside of the body through the penis. It carries semen, and also urine from the bladder, but never at the same time. When a man is going to eject semen, his penis becomes hard and stiff, and the tube from the bladder is shut off. The penis is designed to deposit the semen containing the sperm inside the woman's vagina.

The external organs of the male reproductive system are the penis ***(1)*** *and the scrotum* ***(2)*** *containing the testicles. In each testicle is a mass of seminiferous tubules, where the sperm are formed. The vas deferens loops around the bladder* ***(3),*** *inside the pelvis. The prostate gland* ***(4),*** *below the bladder, secretes seminal fluid with which the sperm mix to form semen. The urethra* ***(5)*** *has two functions; at different times it takes either semen or urine from the bladder through the penis and out of the body.*

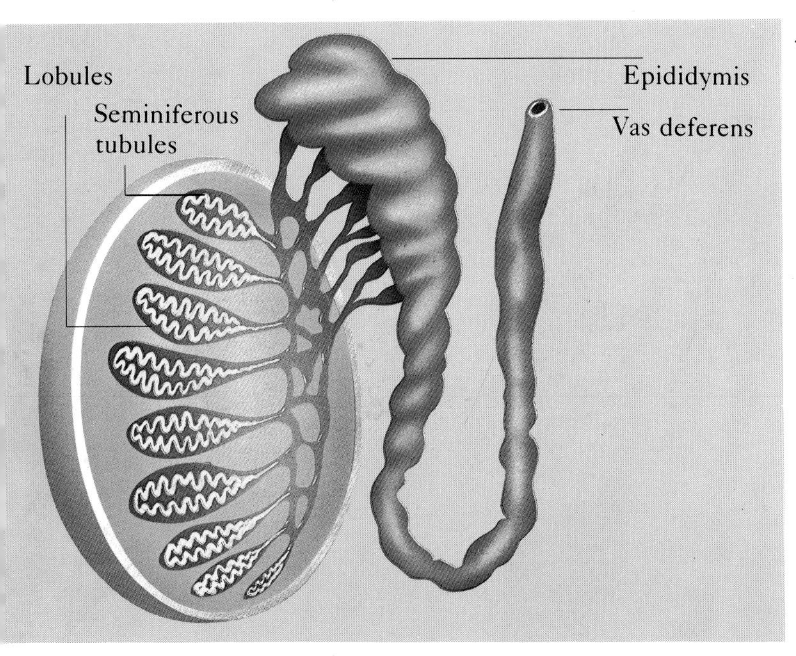

◀ *Inside each testicle is a mass of seminiferous tubules, where the sperm are produced. The network of tubes leads to the epididymis, where the sperm are stored. They leave the body by way of the vas deferens, which connects with the urethra close to the prostate gland.*

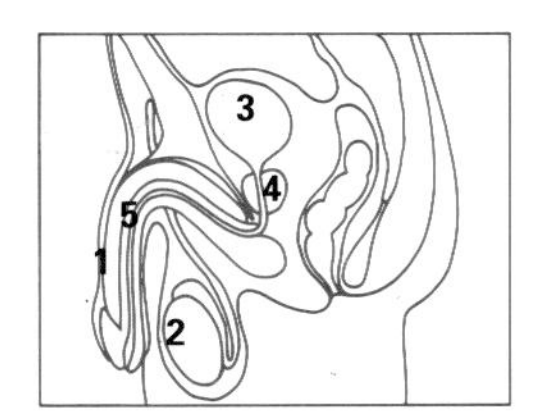

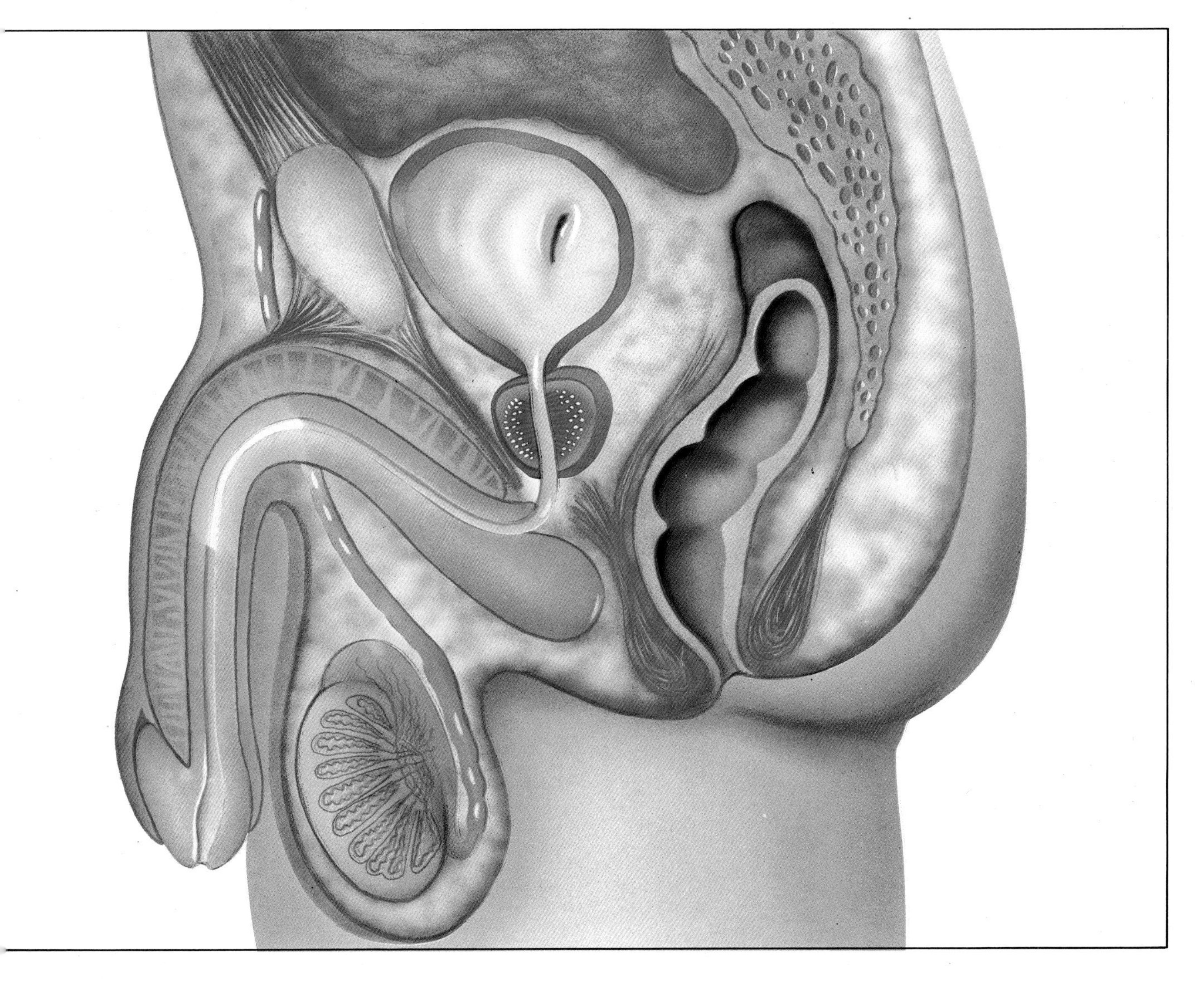

Reaching Puberty

The way in which we grow and develop is controlled by hormones produced in the endocrine glands, which include a female's ovaries and a male's testicles. The endocrine glands produce large amounts of hormones during puberty, and a spurt of growth takes place. Puberty occurs in girls between the ages of about 11 and 14, and in boys a little later, between about 13 and 16.

When a boy reaches puberty, his shoulders and chest broaden. His voice grows deeper, and hair appears under his arms, on his face, and around his penis. His sex organs grow bigger, and his testicles begin to produce sperm.

The most obvious changes that take place in a girl's body during puberty are the growth of her breasts, the growth of hair under her arms and between her legs, and the beginning of menstruation.

Menstruation—also known as the monthly period—is the loss of blood through the vagina. This cycle of blood loss begins about every 28 days and is part of a woman's complicated reproductive cycle.

Each month, an ovum in one of a woman's ovaries matures. This ovum passes out of the ovary into the oviduct and down to the uterus. Meanwhile, hormones produced by the ovary cause the cells lining the uterus—the endometrium—to grow thick with tissue rich in blood vessels. It is getting ready to take in a fertilized ovum.

If the ovum is not fertilized, the hormone production changes. The extra lining of the uterus breaks down and, together with blood from little broken blood vessels, passes out of the body through the vagina. This blood loss lasts for about five days.

A woman's reproductive cycle. The first diagram shows ovulation ***(A),*** *when an ovum that has matured in the ovary passes out of its follicle and into the oviduct. The ovum makes its way along the oviduct* ***(B & C),*** *while the lining of the uterus grows thicker. This is the time of the month when a woman is most fertile. When the ovum reaches the uterus, it may have been fertilized. If it has not been fertilized, the thick lining breaks down and menstruation begins* ***(D).***

▼

A

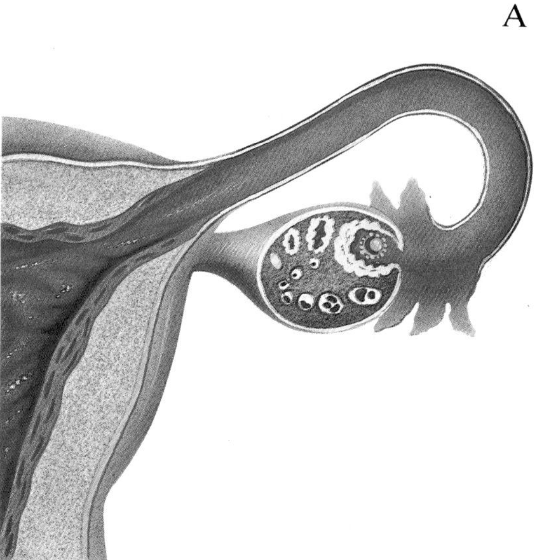

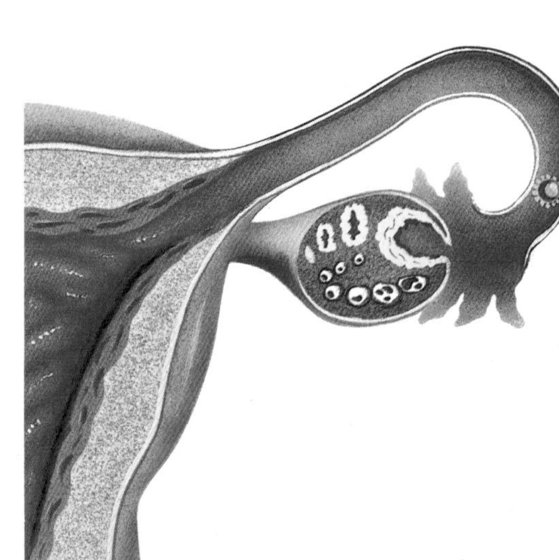

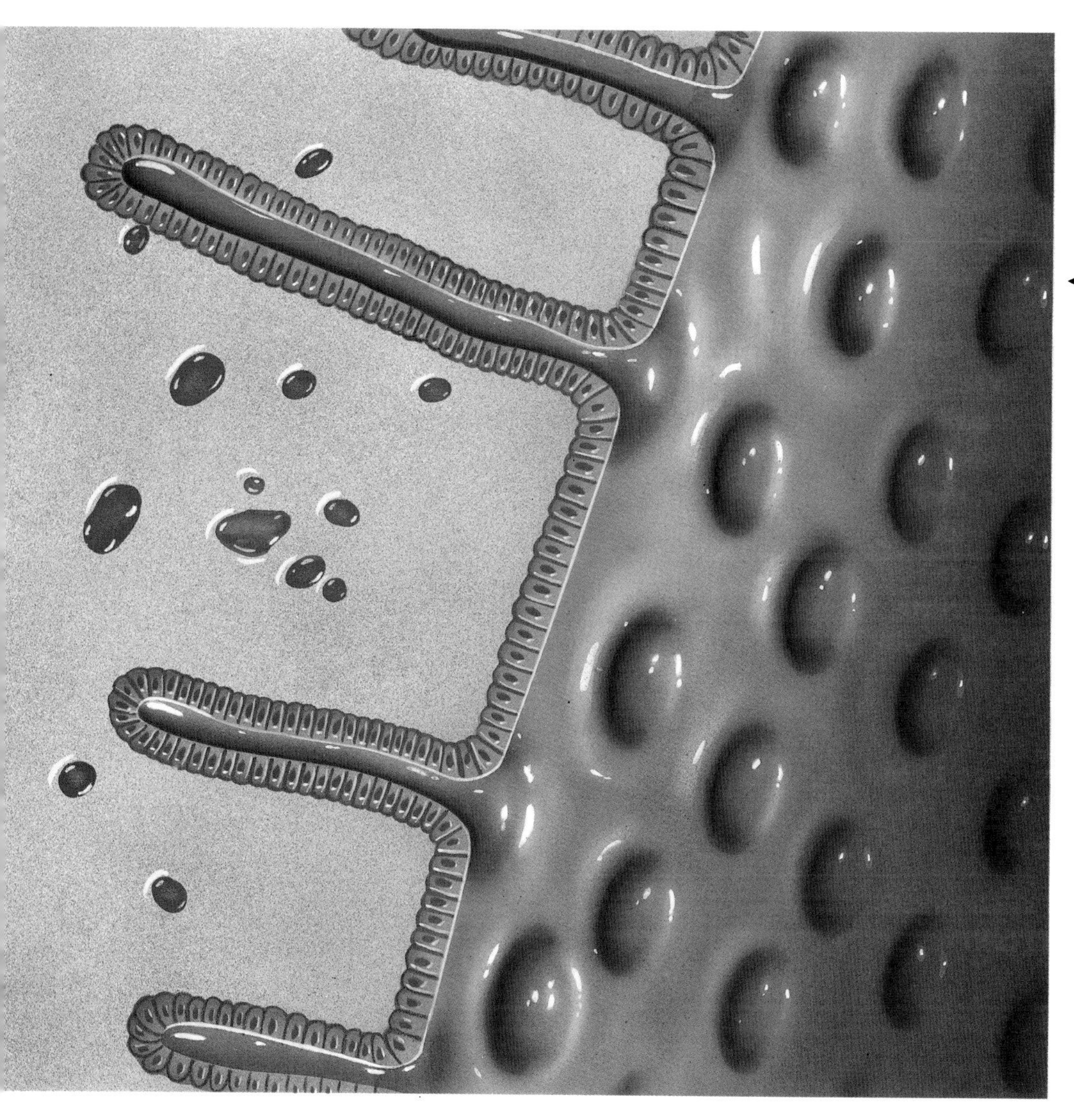

◀ *A close-up of part of the uterus during menstruation. The thickened layer of the endometrium* ***(1)*** *is being shed, because it has no fertilized ovum to look after. Together with some blood* ***(3)*** *from tiny broken blood vessels, it flows out of the body through the vagina. The muscular layer of the uterus* ***(2)*** *remains unchanged.*

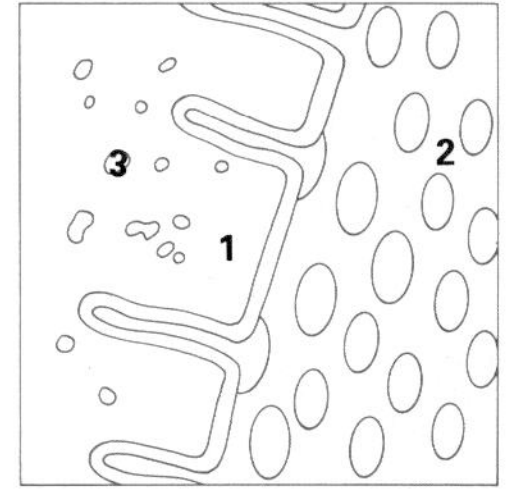

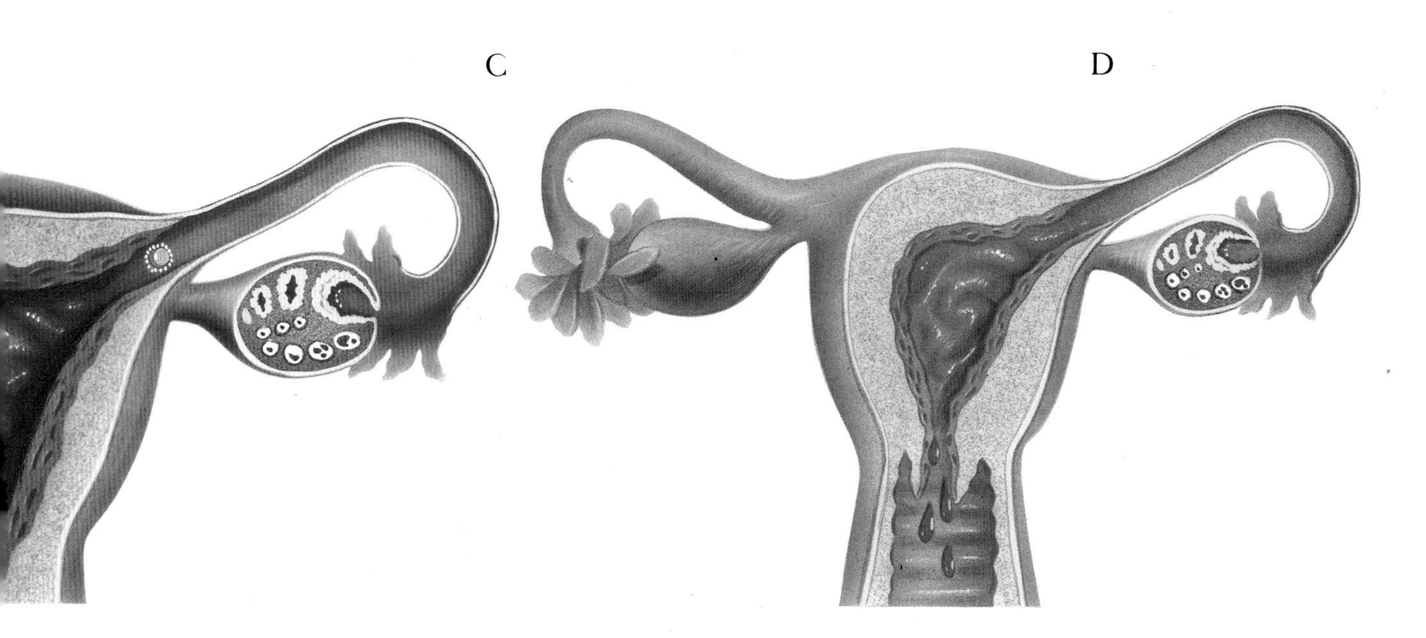

How Ova Are Formed

The female egg cells, the ova, develop inside the ovaries from immature cells. Each ovum is a tiny, specialized cell. It has 23 chromosomes—just half as many as normal human body cells. The only other human cell with 23 chromosomes is the male sex cell, the sperm. When male and female sex cells join together, they form a single cell, or zygote, with 46 chromosomes. The chromosomes carry the genes that hold the genetic information from the parents that will decide how the new person will develop.

An ovum usually lives between 12 and 24 hours. It will live longer only if it is fertilized by a sperm.

When a girl is born, she has hundreds of thousands of immature cells in her ovaries. Most of these will not develop; only about 200 in each ovary will reach maturity and become ova.

In the ovary, each immature cell is surrounded by a round cluster of cells called a follicle. Hormones stimulate the follicles to mature, and about every 28 days between puberty and menopause, a mature follicle breaks and releases an ovum into the oviduct. This process is known as ovulation.

The broken follicle then forms a corpus luteum, or yellow body, that produces the hormone progesterone. If the ovum is not fertilized, within two weeks the corpus luteum shrinks and stops producing progesterone, and menstruation begins. Two weeks later, another ovum matures, and the cycle repeats itself.

A close-up of the inside of an ovary. In the walls of the ovary are clusters of cells called follicles; inside each is an immature egg cell which is growing in size. About every 28 days, a follicle breaks open to release a mature ovum. The broken follicle forms a corpus luteum. This produces progesterone, which gets the uterus ready for a fertilized egg.

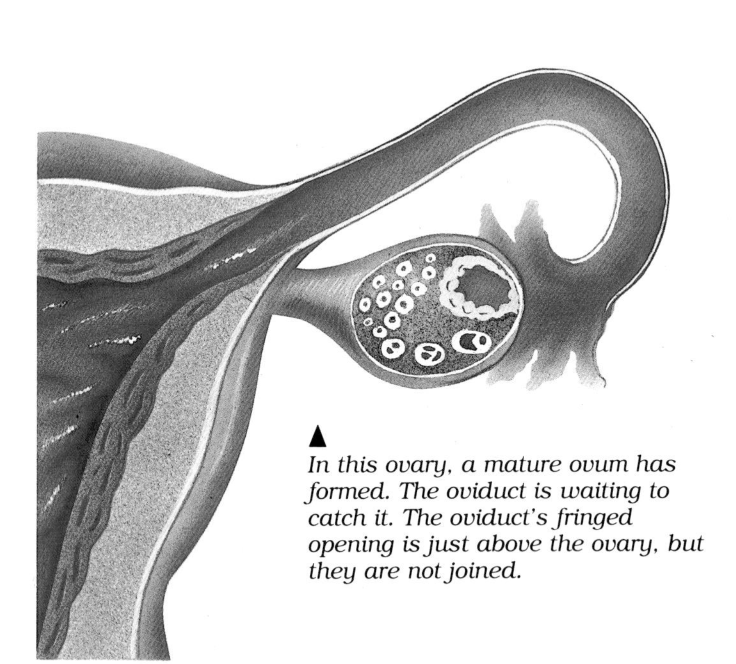

▲ *In this ovary, a mature ovum has formed. The oviduct is waiting to catch it. The oviduct's fringed opening is just above the ovary, but they are not joined.*

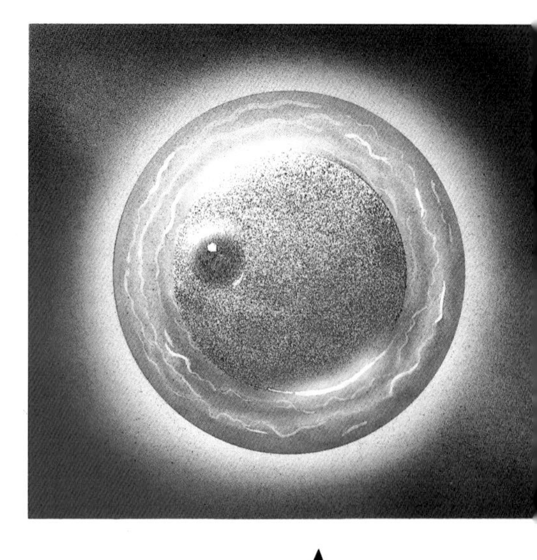

▲ *A mature ovum. Cytoplasm is inside the protective membrane. The round dark spot is the cell's nucleus.*

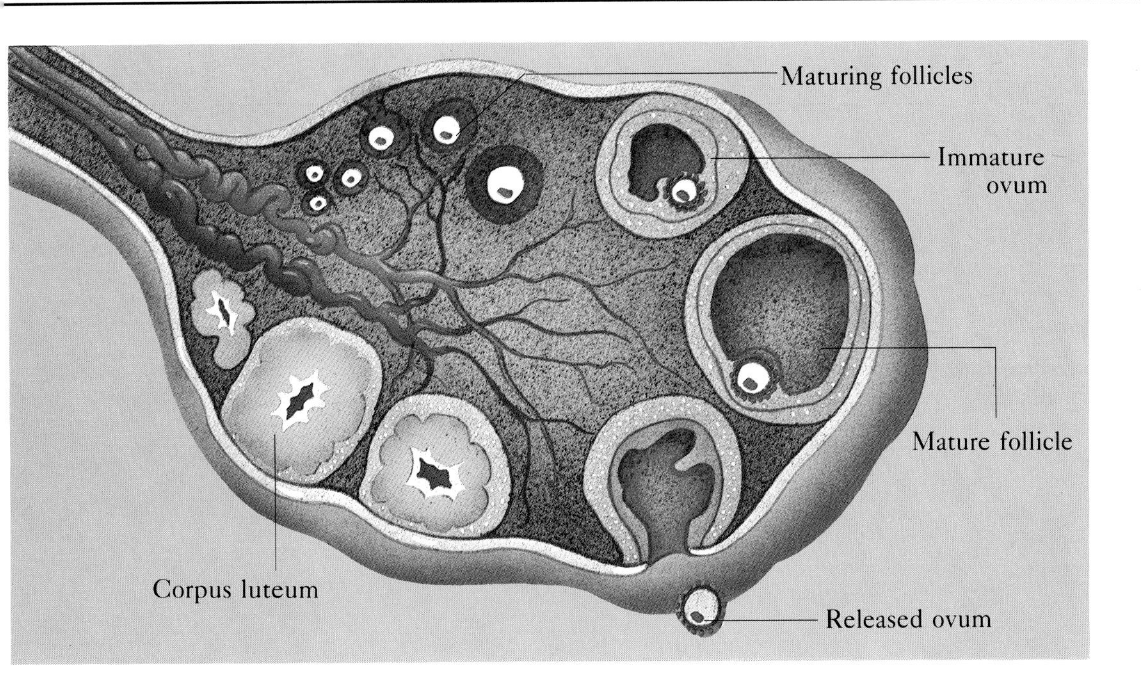

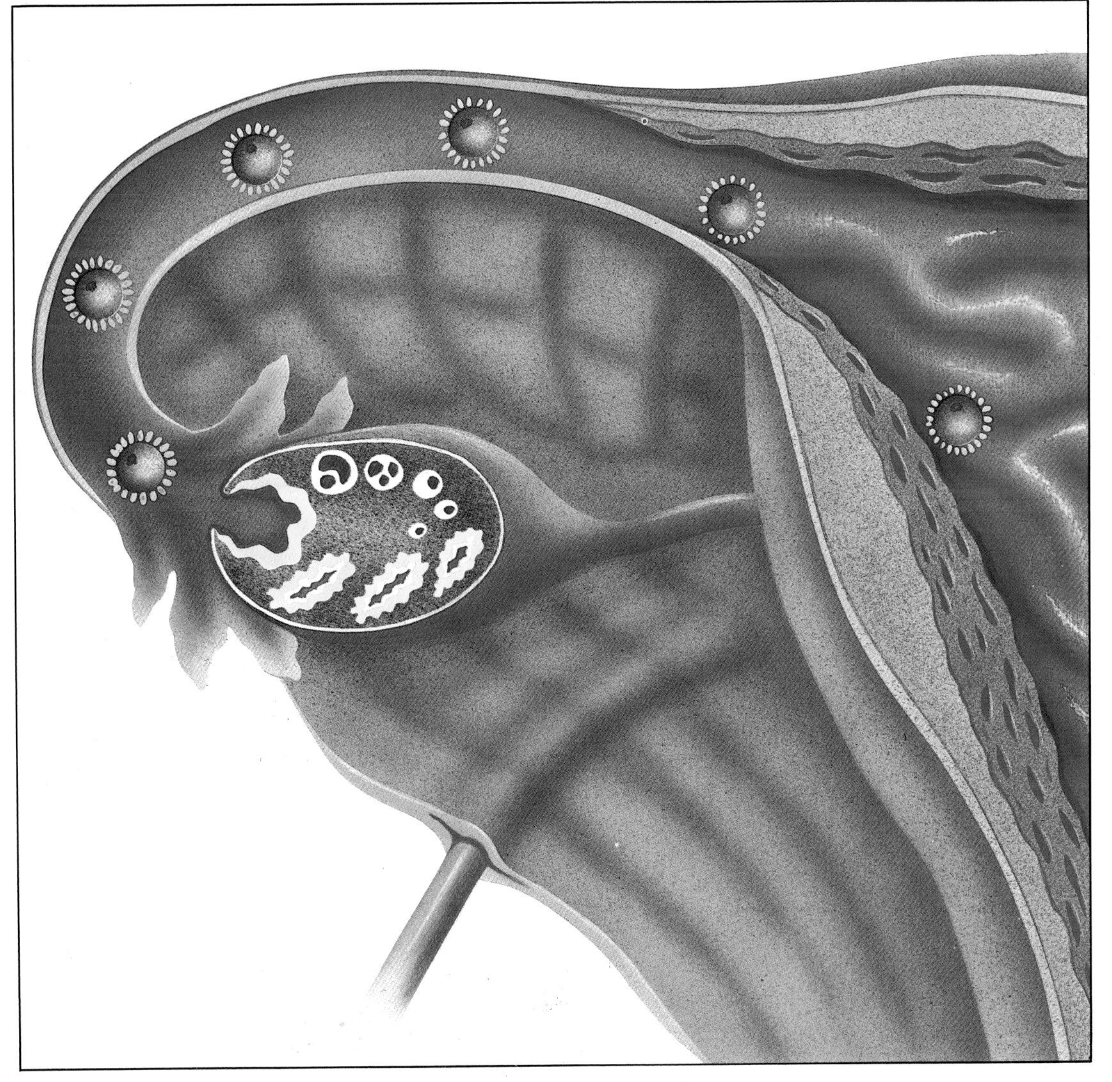

◀ *A mature ovum makes its way from the ovary* ***(1),*** *traveling down the oviduct to reach the uterus, where it may develop. The endometrium, or lining of the uterus* ***(2),*** *has grown thicker, ready to receive a fertilized ovum.*

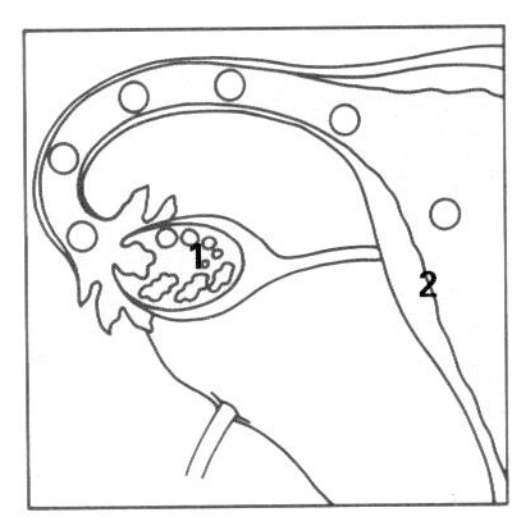

How Sperm Are Formed

The sperm, or male sex cell, is one of the smallest of all human cells and is far too small to see without a microscope. The main part of a sperm is its head, which contains the nucleus, with its 23 chromosomes carrying genetic information. The head is connected by a midsection to a long tail.

Sperm production takes about 40 days. Boys begin to produce sperm during puberty. Production takes place in the testicles, which hang outside the main part of the body so that their temperature is low enough for sperm production.

Inside the testicles, cells with 46 chromosomes divide by a process called meiosis to form new sex cells, each of which has only 23 chromosomes. These cells then develop into the sperm, with the nucleus in the head and the long, mobile tail.

During sexual intercourse, a man releases about 400 million sperm at a time into a woman's vagina. The seminal fluid helps to protect the sperm from acid substances as they make their way up the woman's vagina, through the uterus, and into her oviducts. The sperm move by lashing their tails about.

Most sperm die on their 40-minute journey, and only about 100 arrive in the oviducts. If a sperm meets an ovum, fertilization can take place.

A sperm magnified millions of times. In the head is the nucleus, which contains the chromosomes. At the front is the acrosome, which enables the sperm to penetrate the ovum. The midsection contains mitochondria that create energy for the sperm to move, which it does by thrashing its tail. ▼

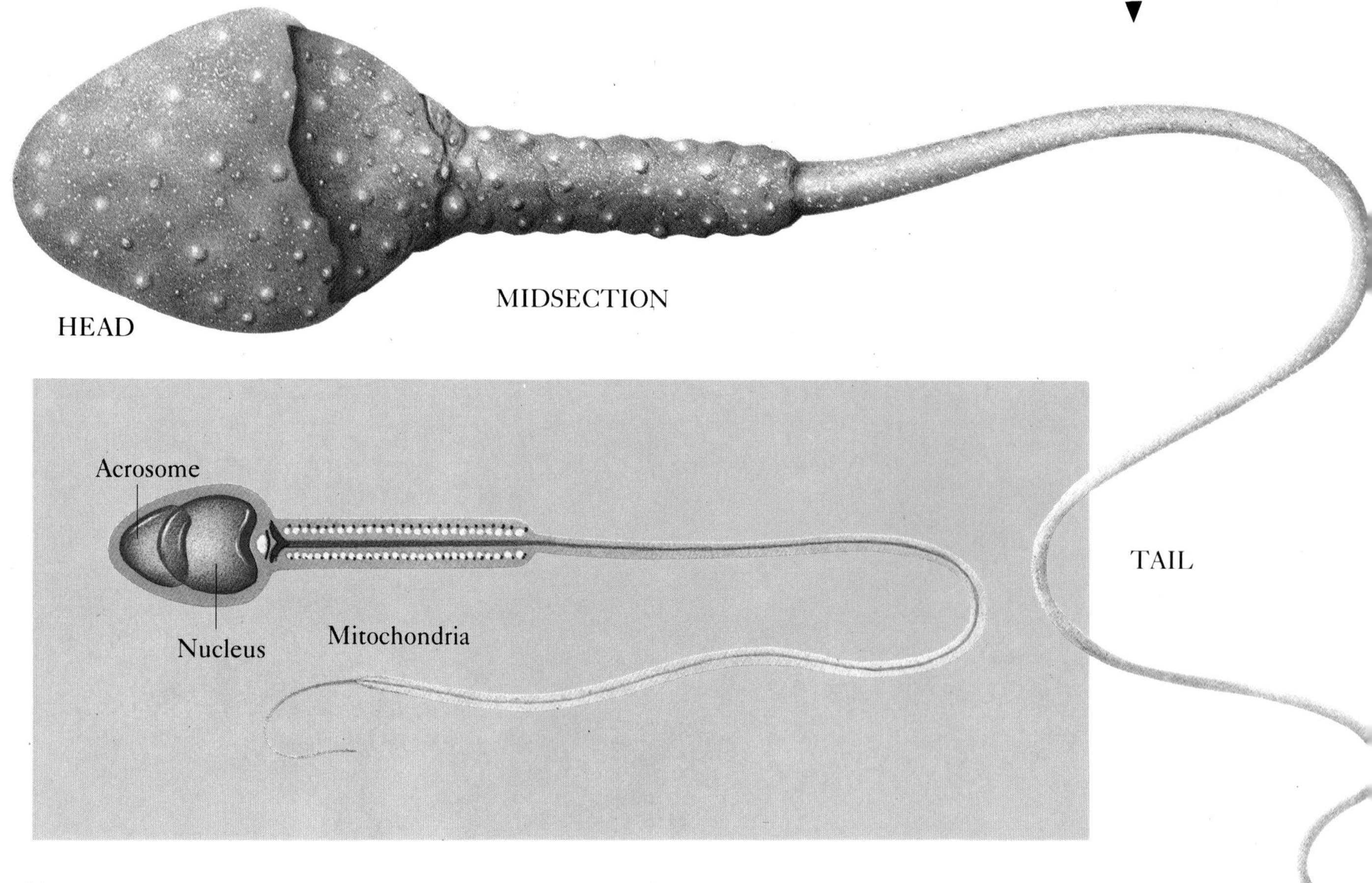

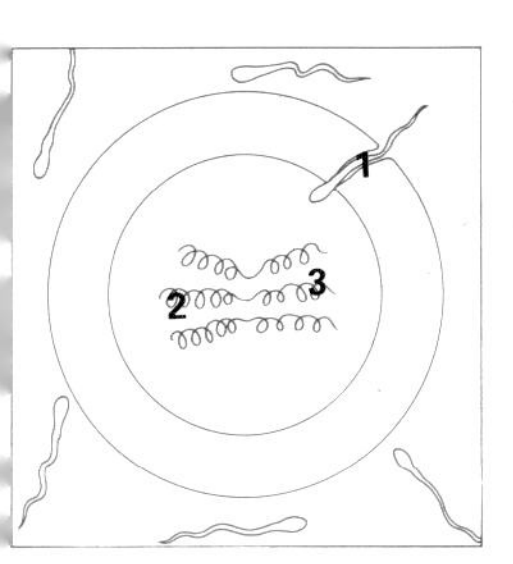

◀ *The sperm* ***(1)*** *has managed to make its way into the ovum, and they join together to form a single new cell. It contains 23 chromosomes from the mother* ***(2)*** *and 23 from the father* ***(3).*** *The two sets of chromosomes pair up, and the information they carry will determine all the new being's physical characteristics.*

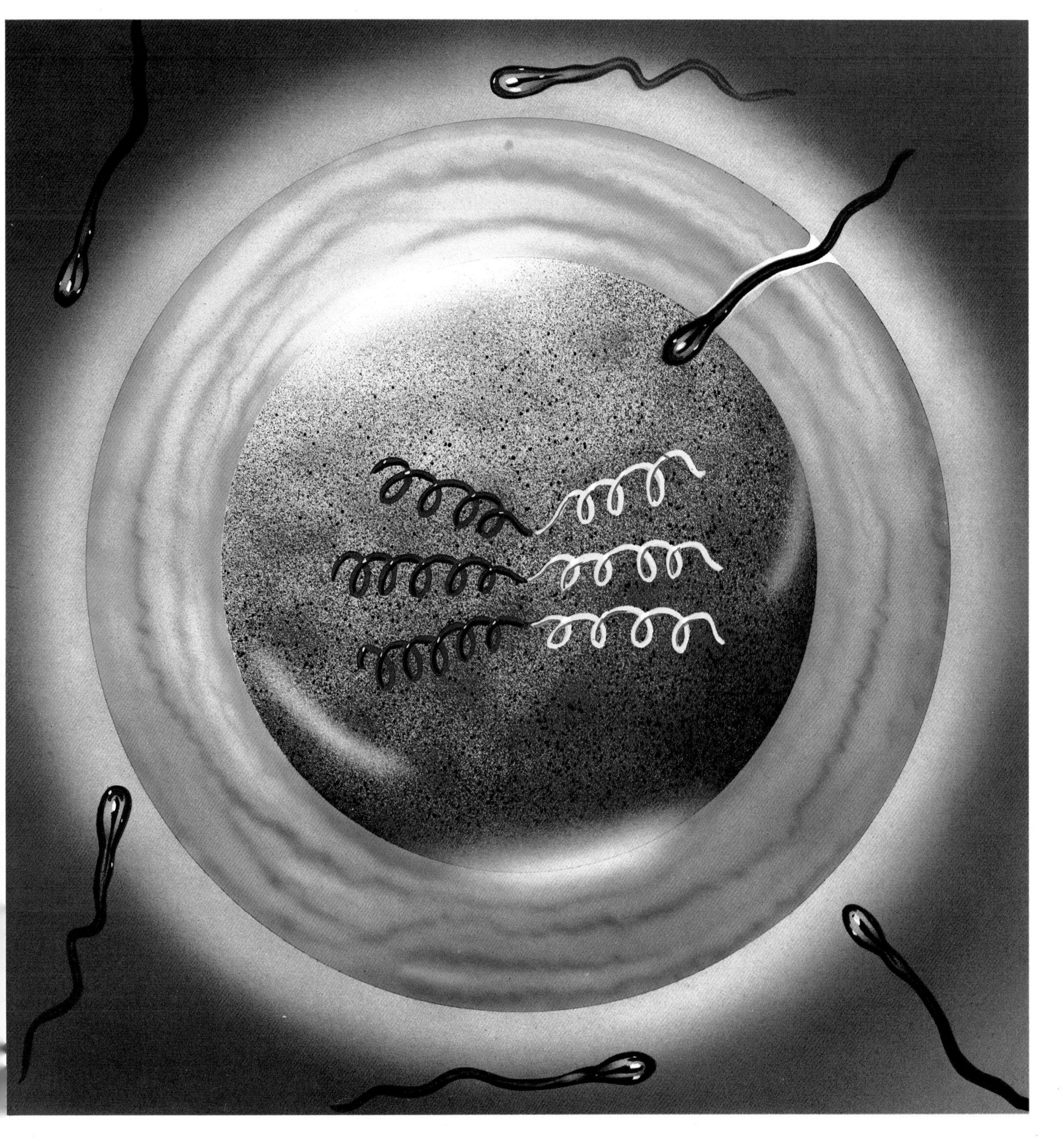

Fertilization

Fertilization takes place when a man's sperm meets a woman's ovum. An ovum can live for between 12 and 24 hours, and a sperm for about three days. Fertilization occurs when the head of a sperm makes its way through the outer surface of the ovum. As soon as this happens, the ovum's membrane changes, and no other sperm can make its way into the ovum.

The new cell formed from the ovum and sperm is called a zygote. In it, the 23 chromosomes from the nucleus of the father's sperm join up with the 23 chromosomes from the mother's ovum to make a cell with a single nucleus containing the normal human cell number of 46 chromosomes. This is the first cell of the new being.

A few hours after the zygote has been fertilized, it divides to make two identical new cells, in a process called mitosis. Then each of the new cells divides into two more, forming four cells. They divide to make eight cells, and so on. Each cell contains exact copies of the chromosomes in the ones before, so the information stored in each cell is identical.

This cluster of cells takes about four days to move down through the oviduct to the uterus.

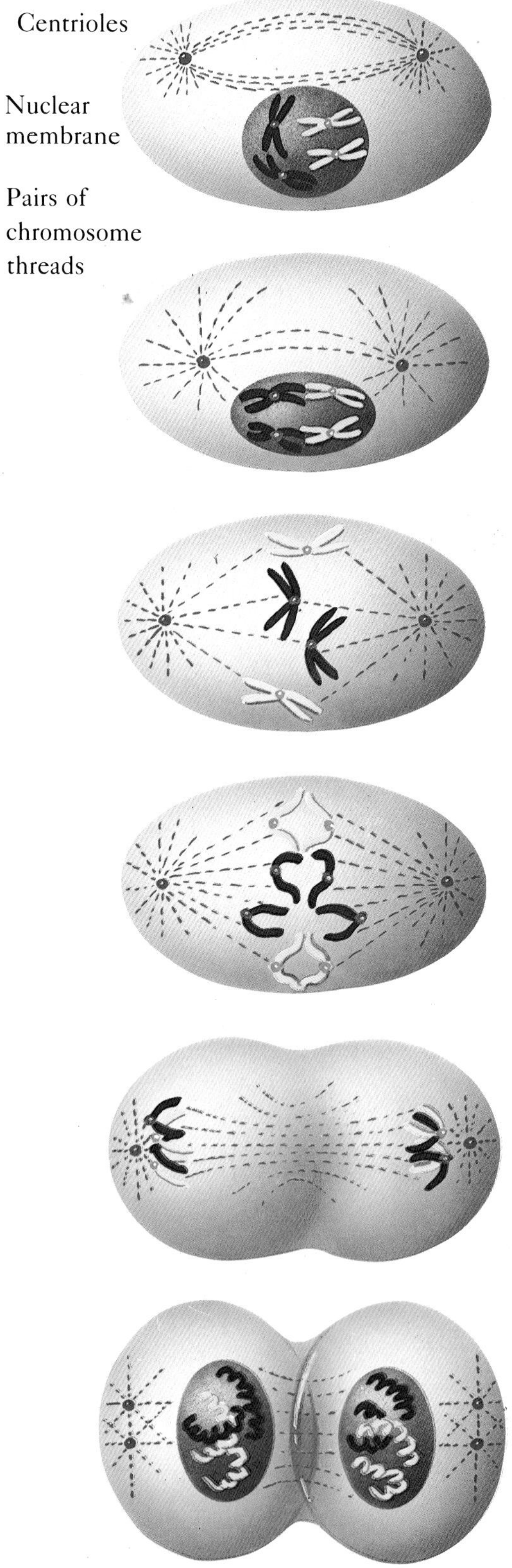

Centrioles

Nuclear membrane

Pairs of chromosome threads

◄ *Division starts when the membrane around the cell's nucleus disappears. The chromosomes move to the center. Structures called centrioles move to opposite sides of the cell.*

The centrioles send out fibers that become attached to the middle of the chromosomes, which then split into two identical threads.

One thread from each moves to each centriole.

A new nuclear membrane forms around each chromosome group. Two new centrioles are formed; the cell divides into two cells exactly like the original.

'he single-celled ygote divides over :nd over again into ', 4, 8, 16, 32, 64 dentical cells, and o on. These are lustered together n a ball, known as a morula. ►

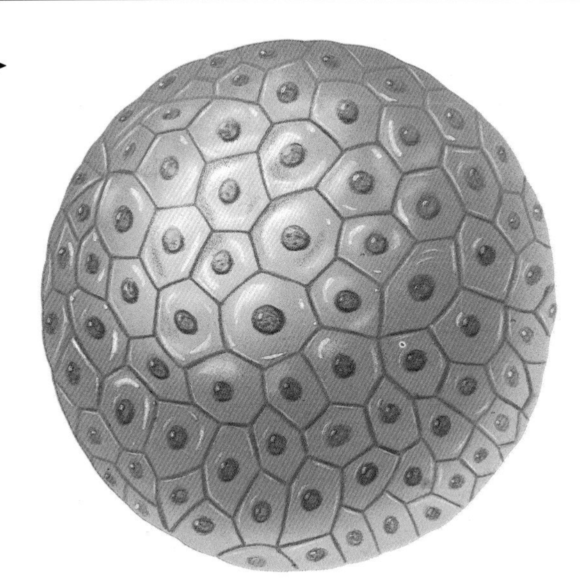

A mature ovum leaves the ovary **(1).** *On its way down the oviduct, a sperm fertilizes it* **(2).** *The egg and sperm combine, and the new cell divides over and over again to form a cluster of cells* **(3),** *which moves along the tube toward the uterus. The lining of the uterus, the endometrium* **(4),** *has thickened; it is ready to receive the fertilized ovum.*
▼

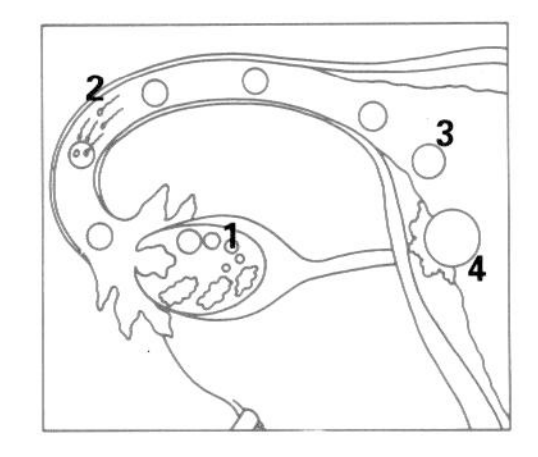

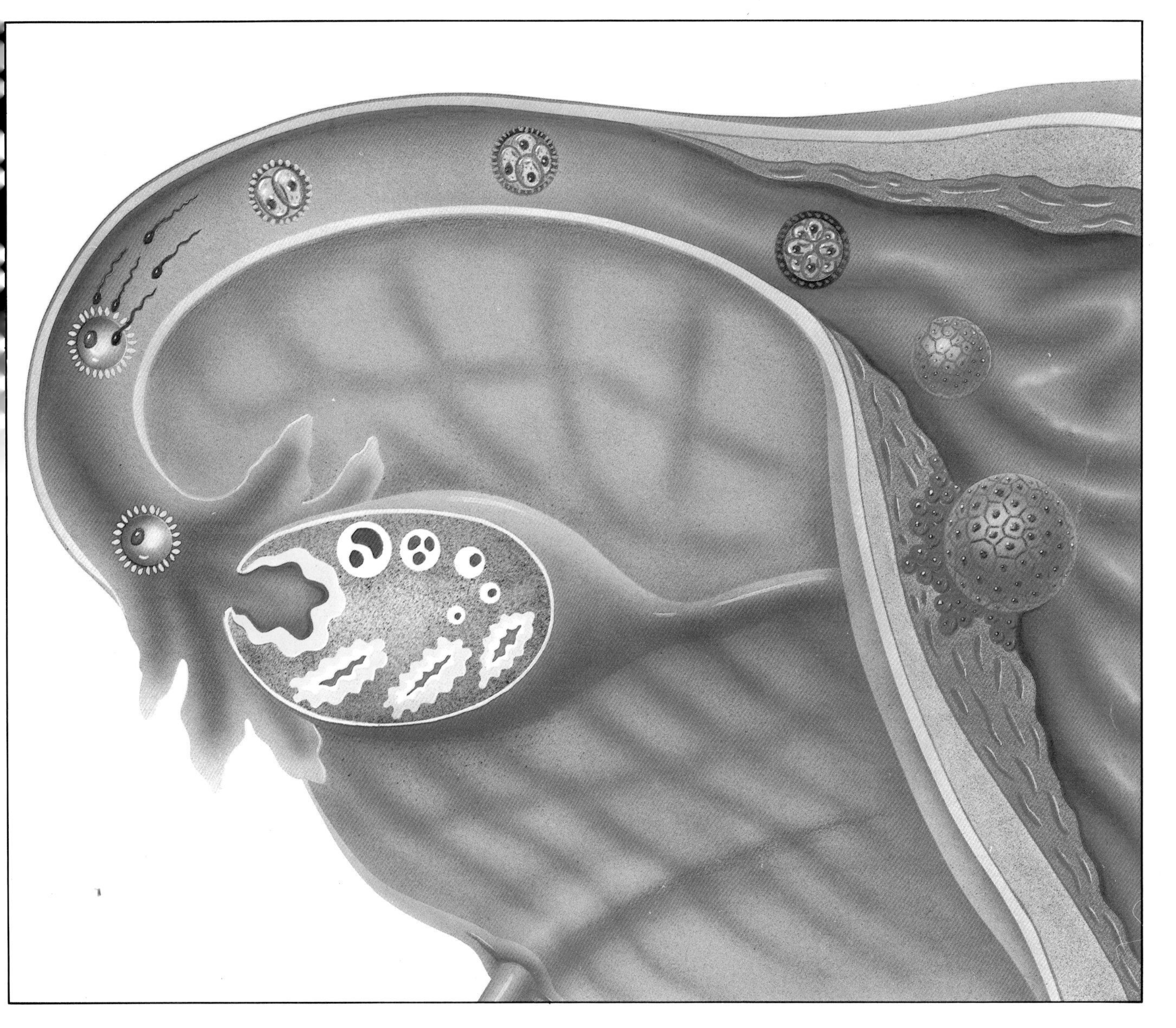

The Uterus

While the cluster of cells has been dividing and traveling down to the uterus, the lining of the uterus—the endometrium—has been changing. It has grown into a thick layer with a rich supply of blood vessels. The ball of cells, known as a morula, implants itself in the lining.

If the endometrium is not ready when the cell cluster reaches it, the cells pass down through the uterus and out of the woman's body. In most cases, however, the endometrium is ready to receive the cluster of cells.

The morula then forms a hollow clump of cells, just big enough to be seen, known as a blastocyst. It forms tiny projections called villi on its outside, which burrow into the endometrium. This stage is called implantation, and from now on the developing cell cluster is known as an embryo. It stays in the uterus, and takes the nourishment and oxygen it needs to grow from the mother's blood supply.

Not all of the dividing cells develop in the same way. Some will become the embryo. Others form the envelopes that surround the embryo, giving it protection and nourishment. These envelopes are the chorion (which produces the placenta) and the amnion.

The zygote divides until it forms a morula. After more cell divisions, it changes into a hollow sphere (shown in the last two pictures) called a blastocyst.
▼

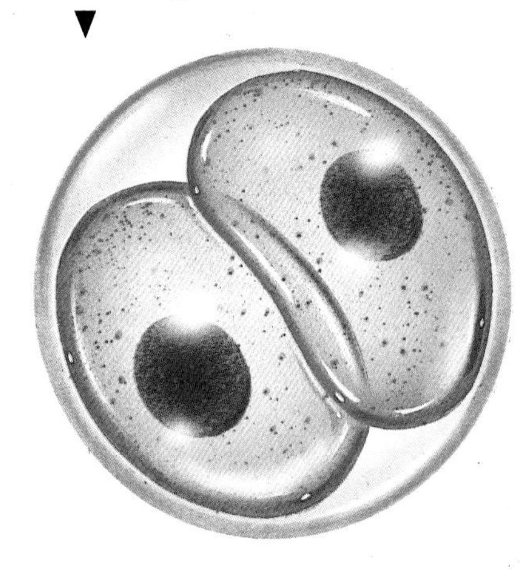
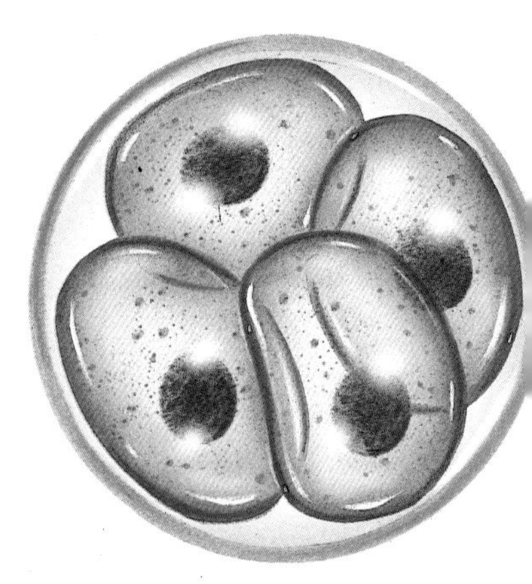
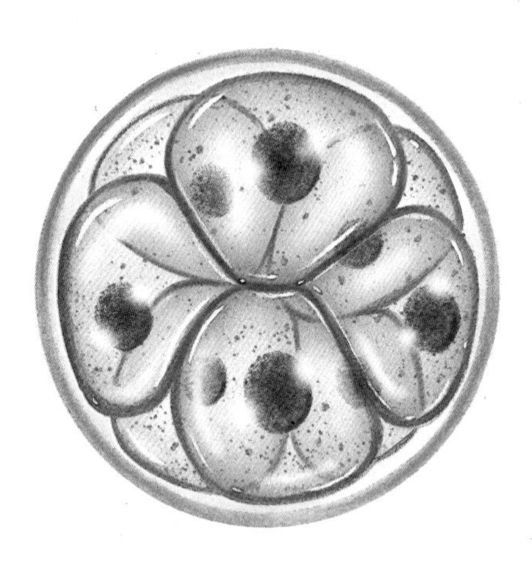
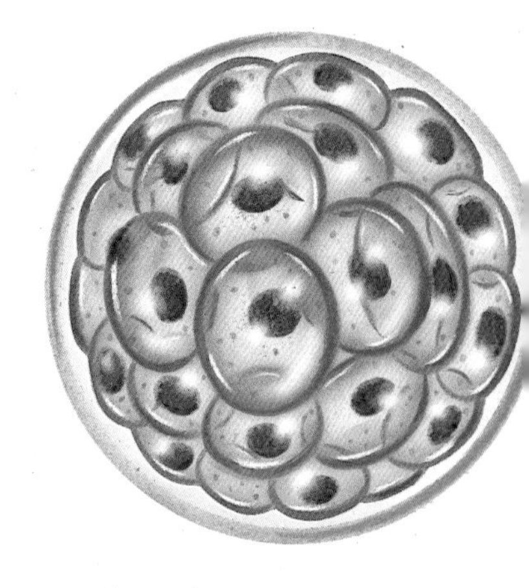
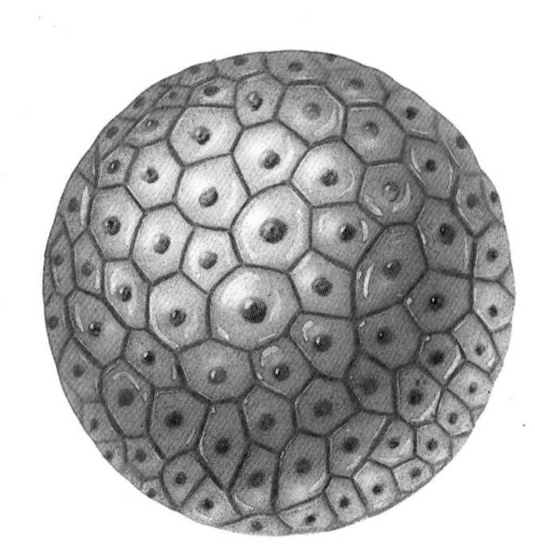
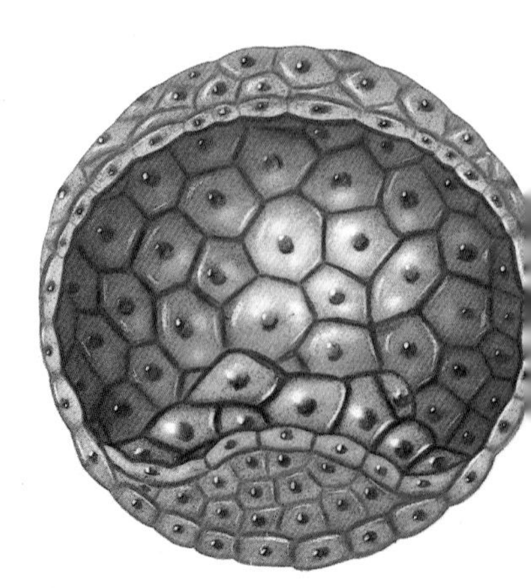

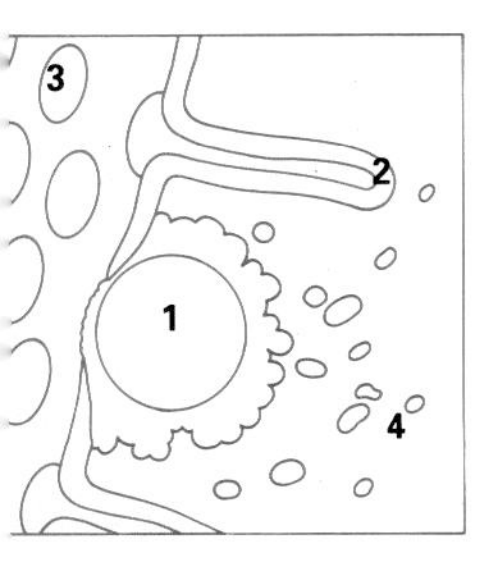

*The blastocyst **(1)** arrives in the uterus **(4)** and "burrows" into the endometrium **(2),** which has developed a layer rich in blood vessels to receive it. The uterus is normally about the size of a small pear; it has muscular walls **(3)** and is hollow inside. Its muscles are so elastic that they can stretch during a woman's pregnancy to allow room for the growing fetus. After the baby is born, the muscles contract and the uterus goes back to its normal size.*

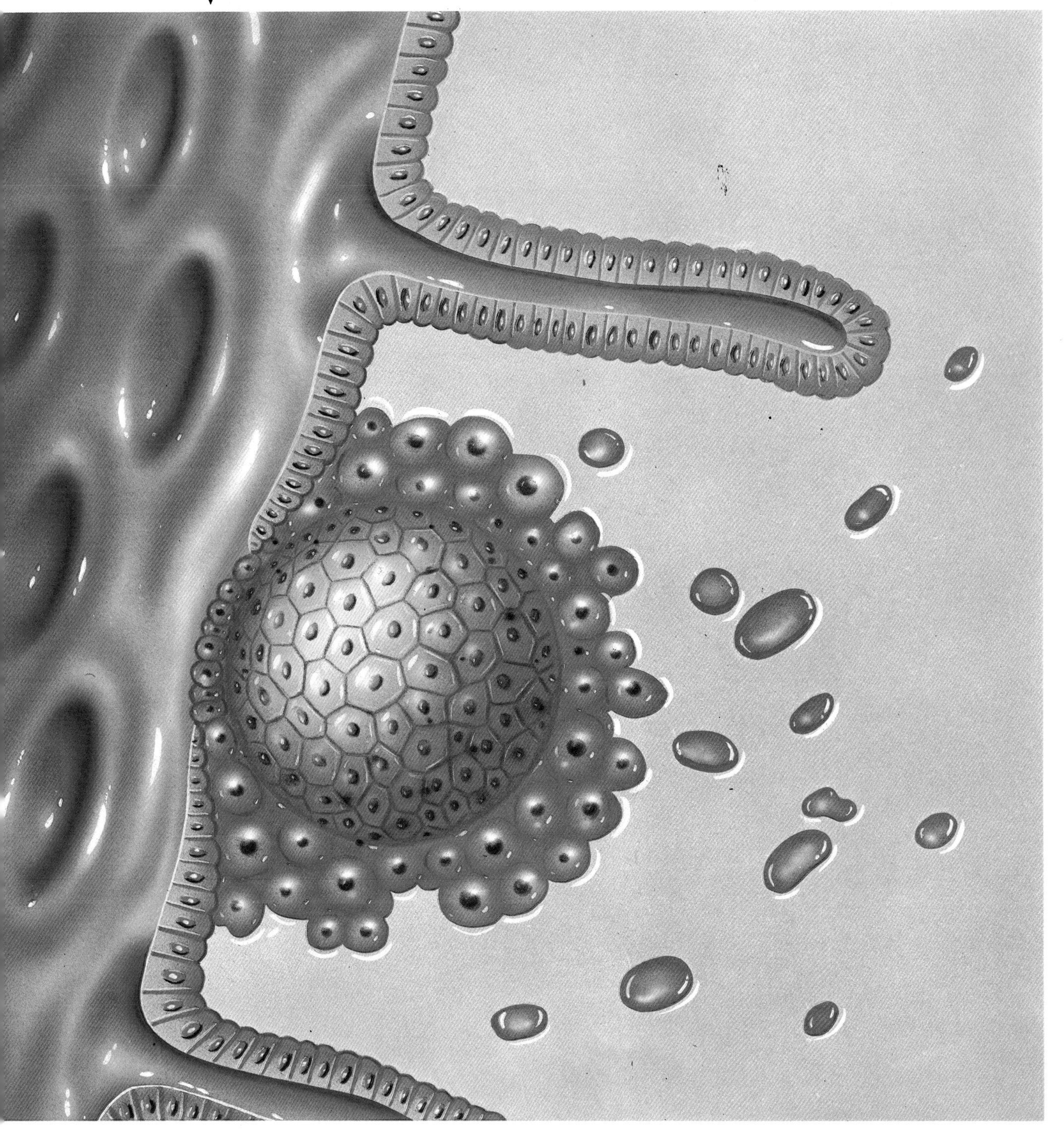

Differentiation

During the second and third weeks of the reproduction process, the embryo settles into place in the mother's uterus. Projections called chorionic villi make their way into the lining of the uterus and begin to take nourishment for the embryo from the mother's bloodstream. These villi develop into the placenta, which is connected to the embryo by the umbilical cord.

A membrane called the amnion develops around the embryo. This membrane is like a bag, and it fills with a liquid called amniotic fluid, in which the embryo floats safely. The liquid cushions it from bumps and knocks, and keeps it at a constant temperature of 98.6° fahrenheit (37° centigrade).

Different types of cells appear in the embryo. They will develop into various parts of the body, such as skin, hair, nails, nerves, the brain, the digestive system, blood vessels, muscles, and bones. As differentiation occurs, the embryo begins to change shape.

After three weeks, the embryo is nearly three millimeters long—about the size of a grain of rice. From then on, it grows very quickly.

By 28 days, the embryo has usually more than doubled its size. The amniotic sac around it is well developed. Its heart starts to form and to beat, at first jerkily but soon with a quick but steady rhythm.

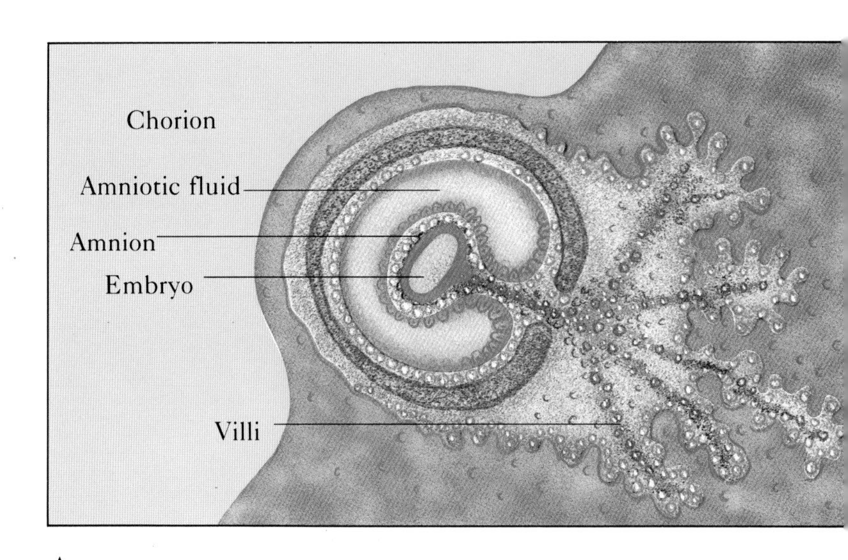

▲ *Finger-like villi reach out from the chorion into the endometrium. They take food for the embryo from the mother's bloodstream.*

The embryo has settled into the endometrium ***(A).*** *By this time, it has three distinct layers of cells: the ectoderm* ***(B),*** *the mesoderm (not visible), and the endoderm* ***(C).*** *It is protected by the amniotic fluid* ***(E),*** *while the chorion* ***(D)*** *absorbs food from the endometrium.* ▼

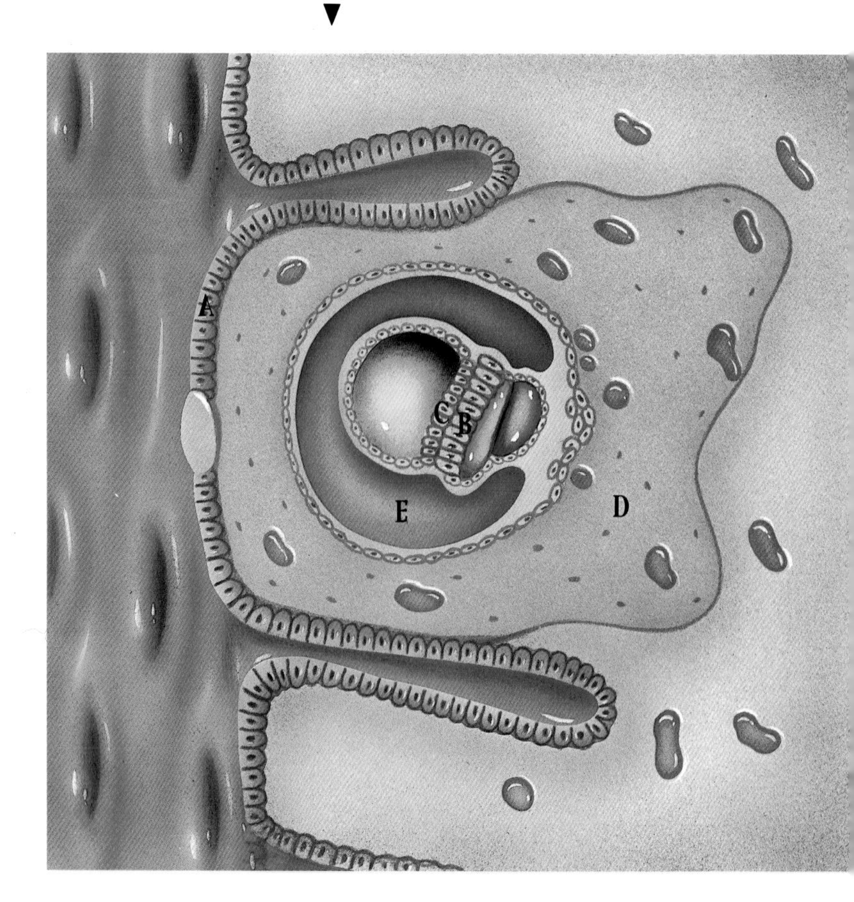

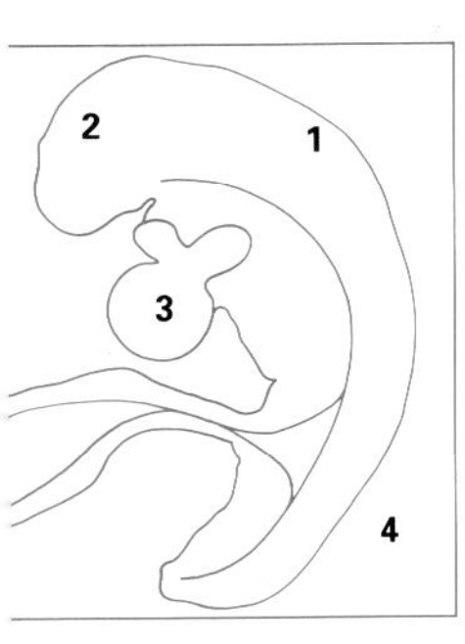

The new being is beginning to take shape. The spinal column ***(1)*** *is already beginning to form, and an arching cluster of cells at one end will develop into the head* ***(2).*** *At this stage, the heart* ***(3)*** *is just a broadening of a blood vessel that looks like a bag. The embryo is wrapped in a membrane called the amnion, which will develop into a pouch around the embryo, full of fluid. This amniotic fluid* ***(4)*** *protects the embryo from bumps and keeps it at a constant temperature. The embryo is connected to the placenta by the umbilical cord.*

▼

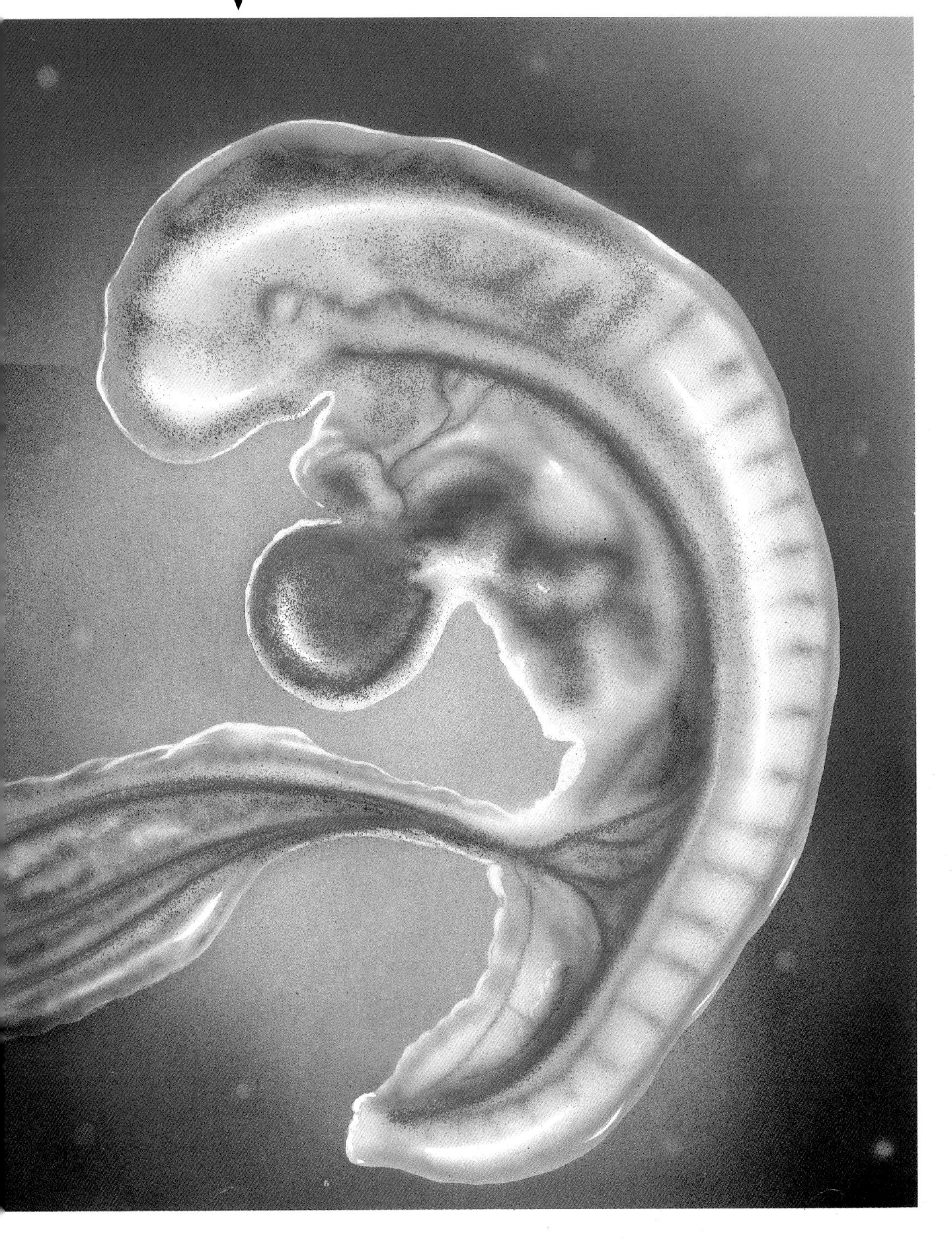

The Body Forms

During the fifth week, the embryo's legs and arms are forming. At first, they look like buds, but soon they lengthen. In the seventh week, shoulder and elbow, hip and knee joints, and the lungs begin to develop. Hands and then feet form. The fingers and toes are webbed, but by the ninth week they have become separate. Meanwhile, the embryo's digestive system is forming.

At five weeks, the embryo is about 10 millimeters long; at the end of the sixth week, it is 13 millimeters long. By the the neck is starting to take shape. The eighth week, it is 40 millimeters long.

By the time two months have passed, the embryo's eyes are covered with a layer of skin which will form its eyelids. Its ears have taken shape, and so has the beginnings of a mouth. The head is bent over the body, but embryo is beginning to move its limbs.

At this point, the embryo is known as a fetus.

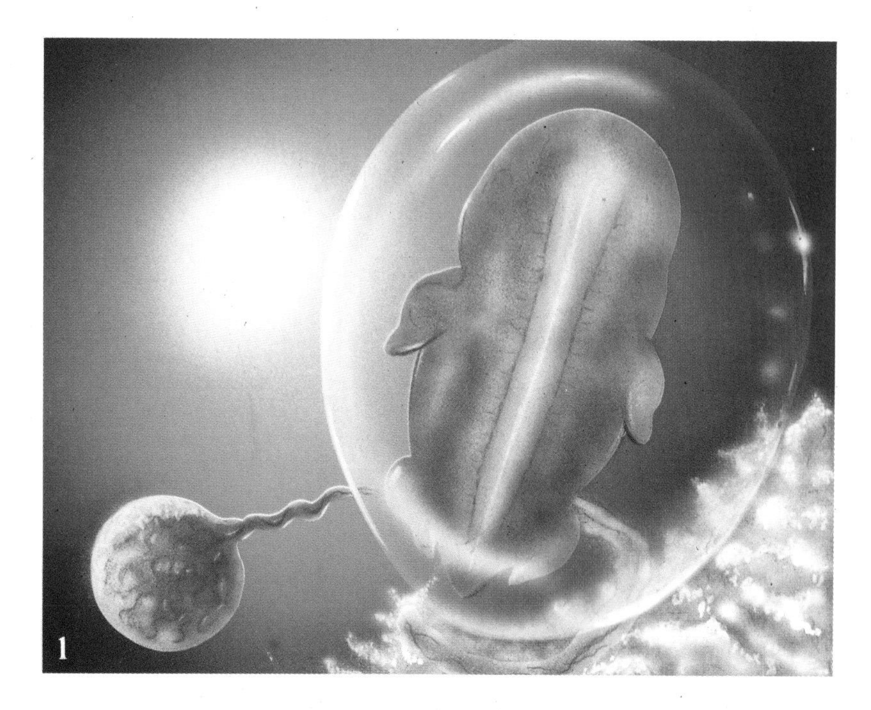

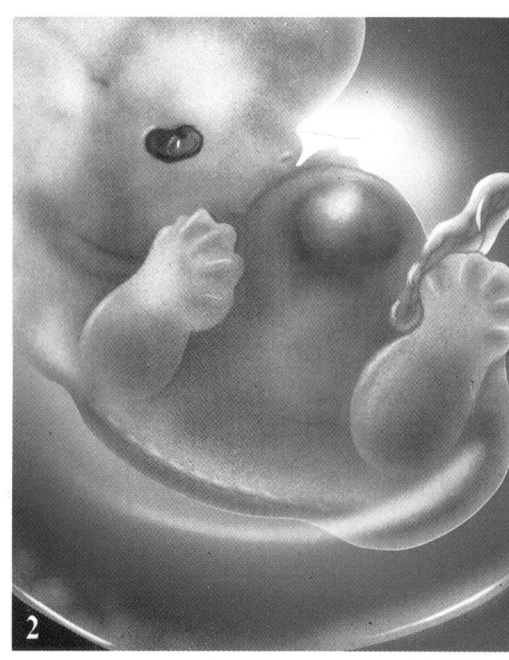

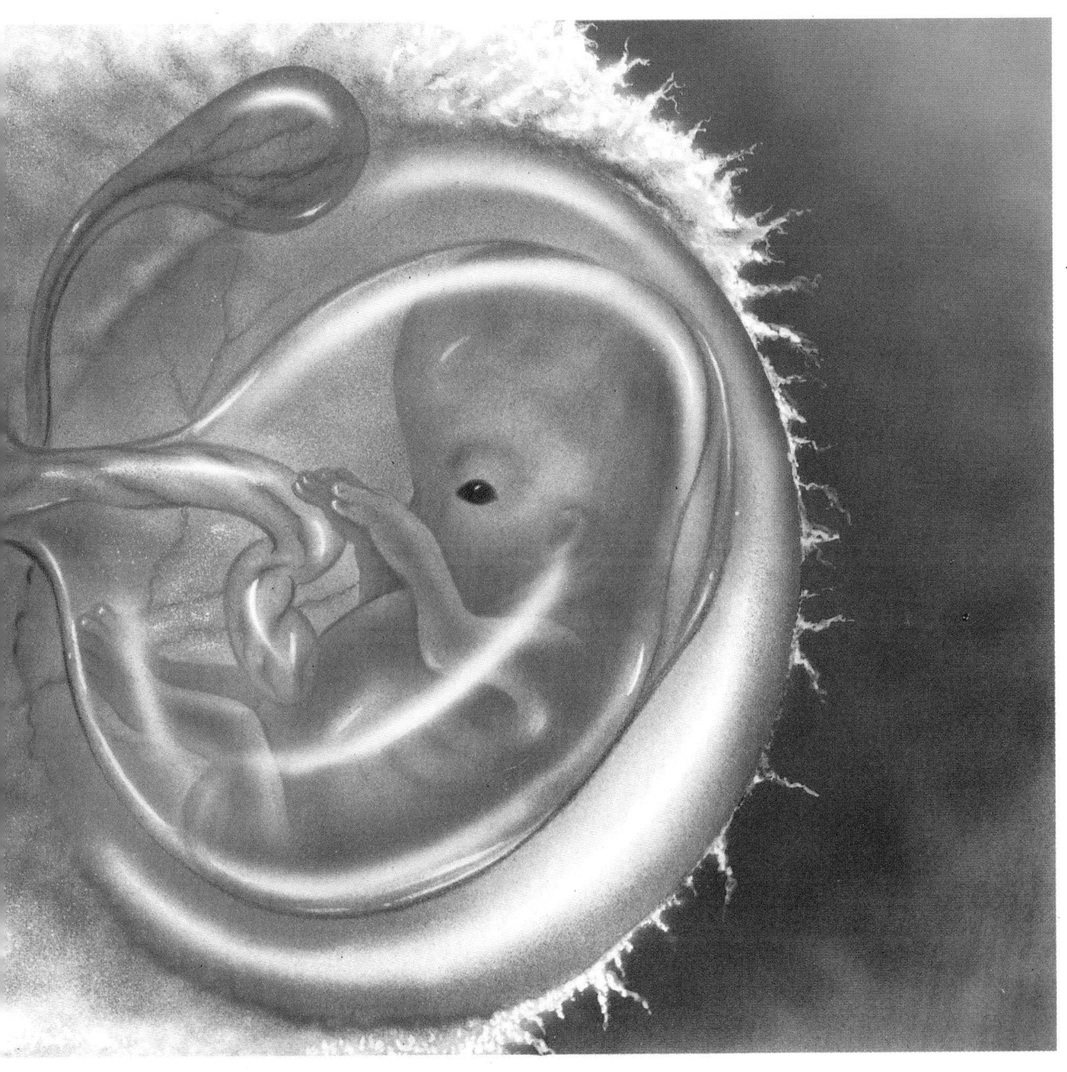

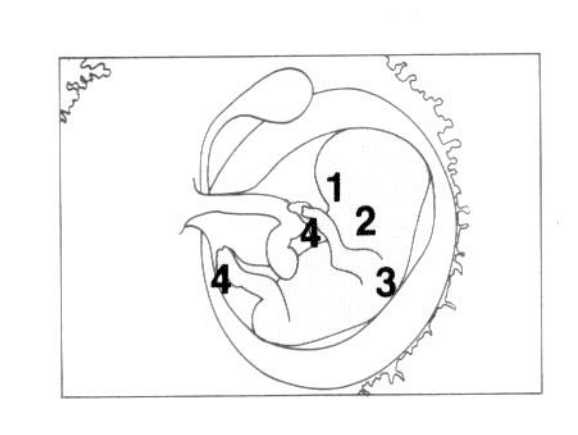

◄ *In the sixth week of the embryo's development, its eyes **(1)** and ears **(2)** begin to take shape. The head starts to stretch out from the body, and the neck **(3)** becomes defined. The hands and feet **(4)** begin to form.*

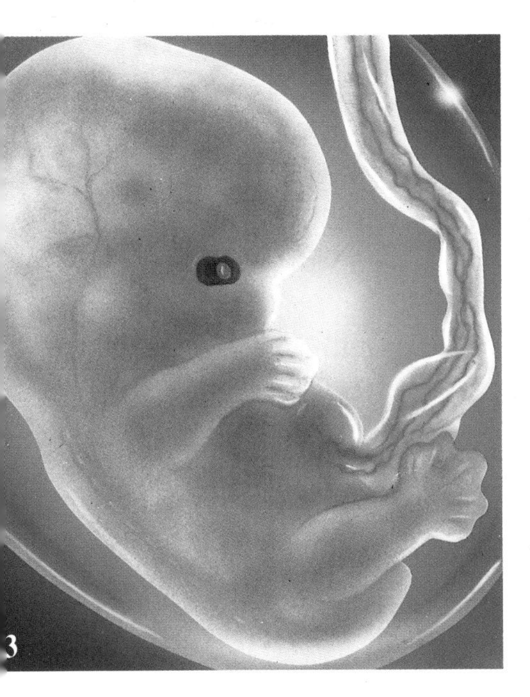

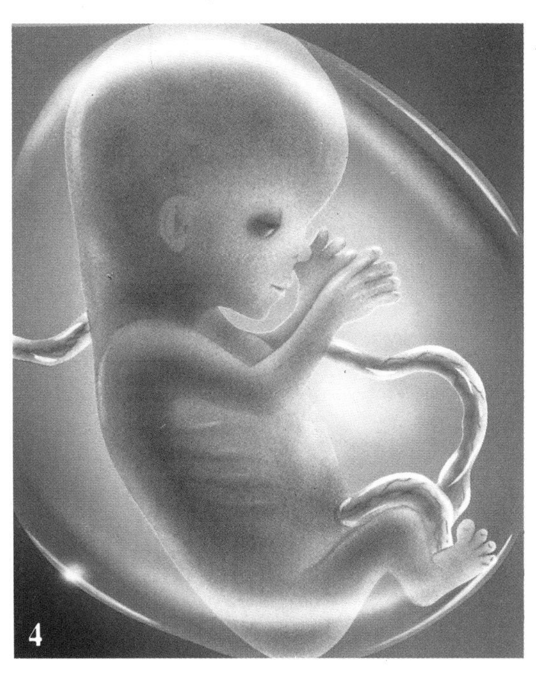

◄ *The embryo grows quickly inside the amniotic sac. The head, spine, and limb buds form **(1)**; the head is enormous compared with the rest of the body and already has eyes **(2)**. Limbs develop joints and soon grow fingers and toes **(3)**. After two months, the fetus begins to look like a baby **(4)**.*

The Fetus Grows

Until the end of the second month, the embryo has fitted into the hollow center of the mother's uterus. From now on, the uterus has to stretch to make room for the fetus, its amniotic sac, and the placenta. The mother's abdomen will soon begin to swell out, and her pregnancy will show more and more.

During the next weeks, the fetus's inner organs grow. Muscles start to form, and nails and hair grow. It moves its limbs around increasingly strongly.

In about week 12, the mother may see an ultrasound picture of her baby, built up through high frequency sound waves and shown on a screen. This picture allows doctors to make sure that the baby is developing properly. They can also usually tell whether the baby will be a boy or a girl.

By week 14, the fetus is fully formed, though very tiny; it is 5 inches (12 centimeters) long, and weighs less than 5 ounces (135 grams).

By week 20, the fetus is moving around so strongly that the mother can feel its movements. The fetus can open and close its fists and its eyes, it falls asleep and wakes up frequently, and it even has hiccups! If a fetus is born in week 24, it is possible for it to live, even though it weighs less then 2 pounds (1 kilogram) and measures only about 12 inches (30 centimeters) long.

In the fifth month, the fetus moves about in the amniotic fluid, sucking its thumb and kicking so strongly that the mother feels the movements.

This six-month old fetus is floating in the amniotic fluid ***(1).*** *The umbilical cord* ***(2)*** *is the fetus's only source of food and oxygen.*

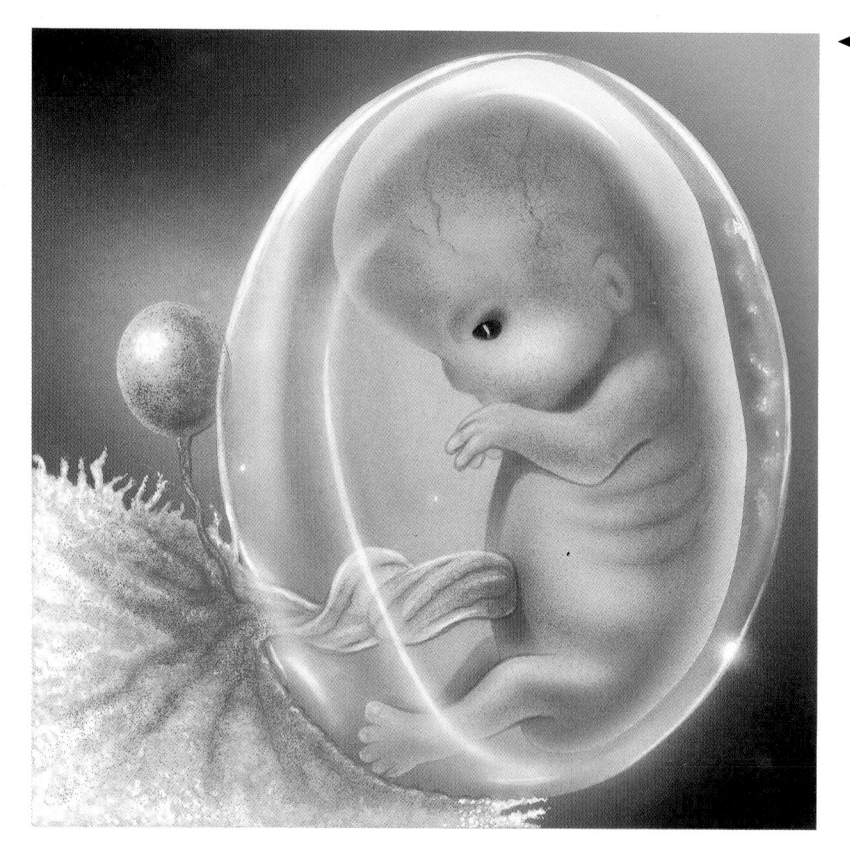

◀ *During the early weeks of a fetus's development, its head is very large in proportion to the rest of its body. The eye sockets develop in the fifth week; they are soon covered by skin, which will form the eyelids.*

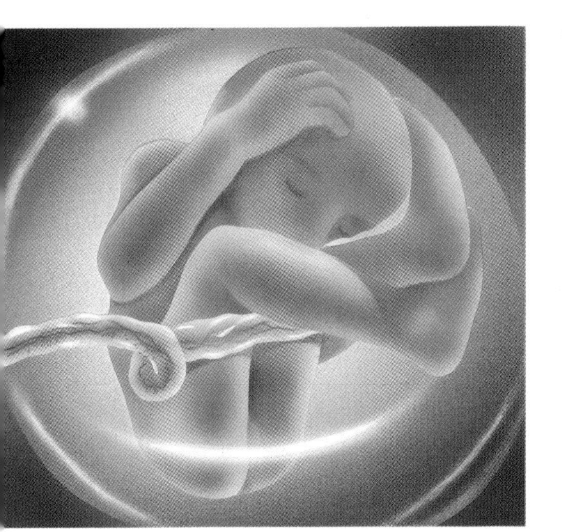

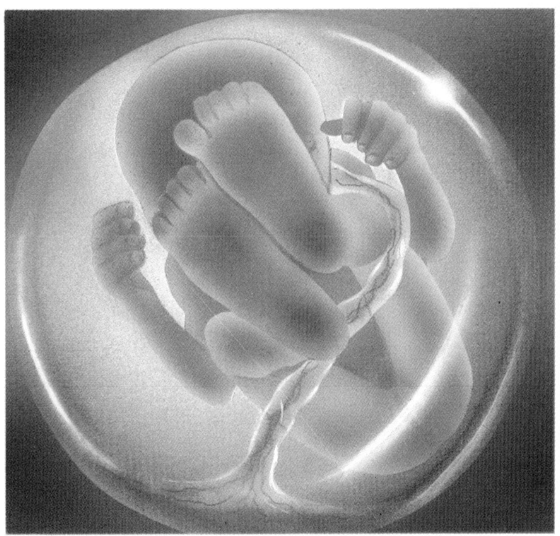

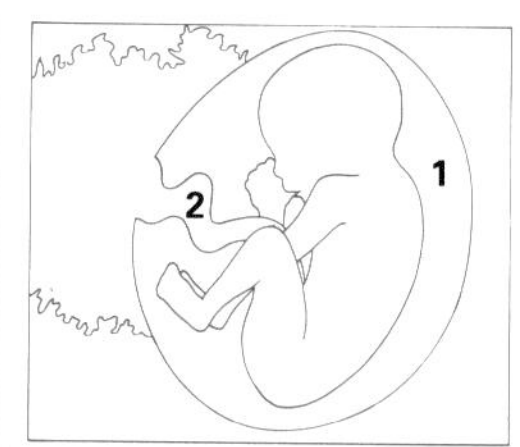
1
2

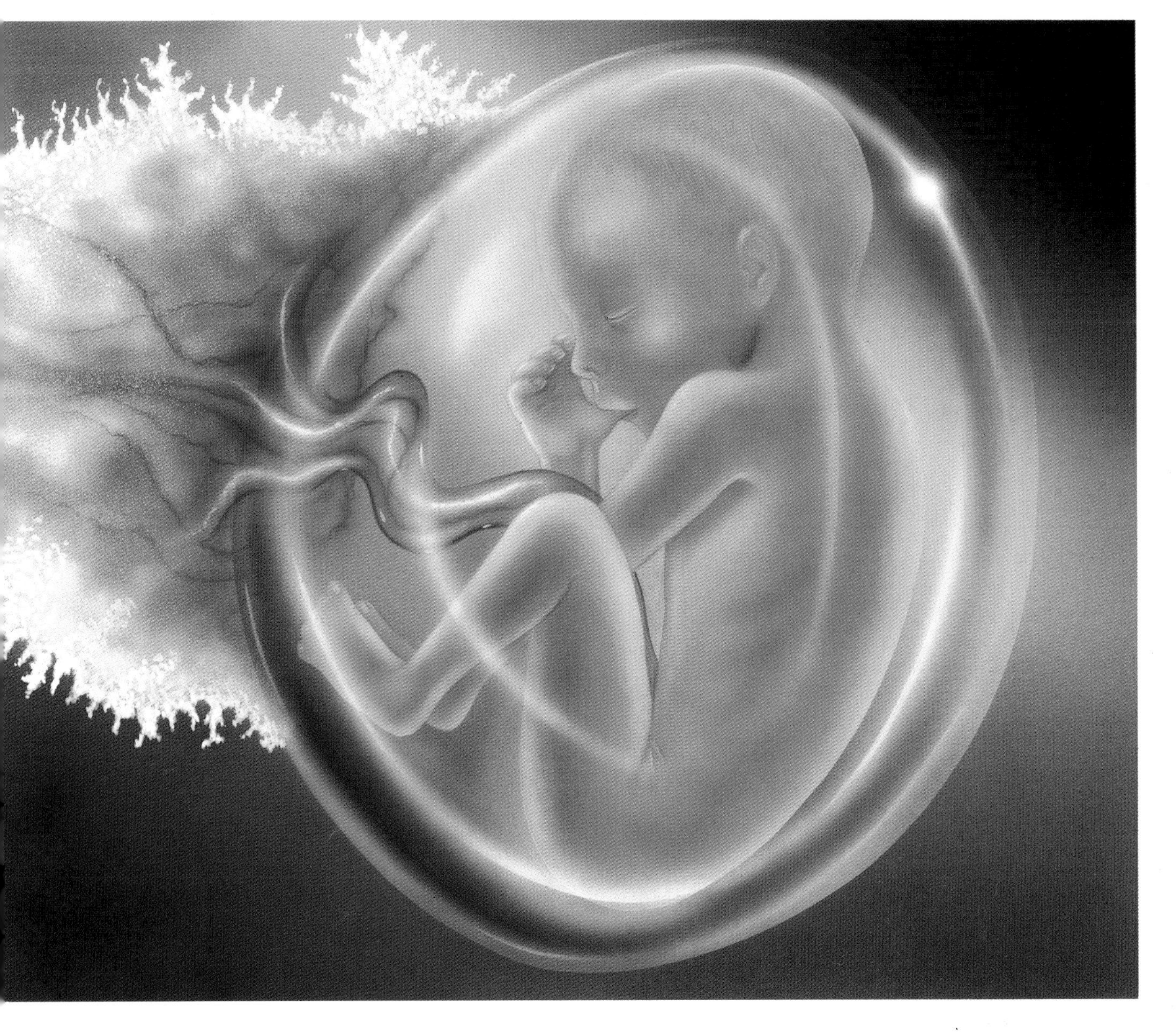

The Placenta

The development first of the embryo and then of the fetus is made possible by the placenta, the vital link between the new being and its mother.
The placenta develops from the little projections that grow from the embryo into the lining of its mother's uterus, and from blood vessels in the lining. The blood vessels from the embryo (and later, the fetus) and from the mother are separated by two layers of cells; between them is a space filled with the mother's blood.
The placenta is connected with the fetus by the umbilical cord, which joins the fetus at the abdomen. A vein and two arteries run through the cord. The umbilical arteries take fresh blood carrying oxygen and nutrients from the mother to the embryo; the umbilical vein carries blood with carbon dioxide and the fetus's waste products to the placenta.
After 38 weeks, the fetus is usually ready to be born. In the first stage of labor (as the birth process is called), the mother begins to have strong, regular muscle spasms called contractions. Gradually, the cervix, the lower end of her uterus, opens wider and wider. After about 12 hours (this time can vary quite a lot), the cervix is some 10 centimeters in diameter. The second stage begins, and it usually lasts for about an hour. Even stronger contractions push the baby through the cervix, down the vagina, and out of the mother's body. The umbilical cord is cut and tied to form the baby's navel. More contractions push out the placenta. Now the baby is ready to live in the world, though it will need care and help for many years.

Three months after fertilization, the placenta is fully formed. The heart of the fetus pumps blood to the placenta through the umbilical cord. Two layers of cells keep the fetus's blood separate from that of its mother, but allow waste products from the fetal blood to pass out, and oxygen and nutrients from the mother's blood to pass into the fetus's blood.
▼

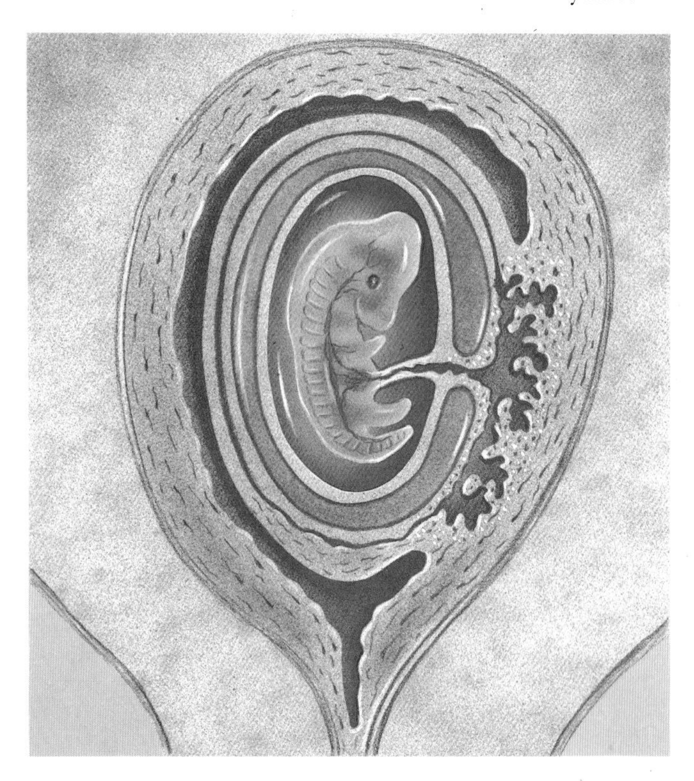

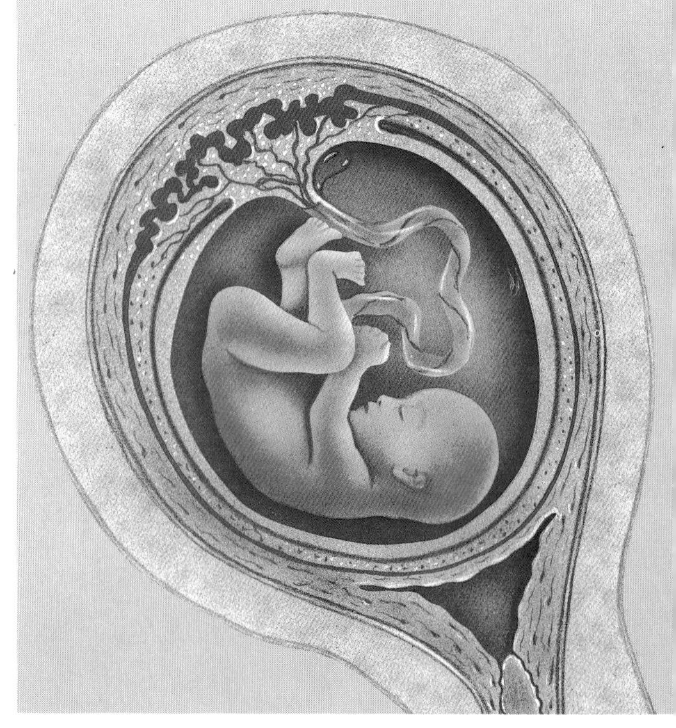

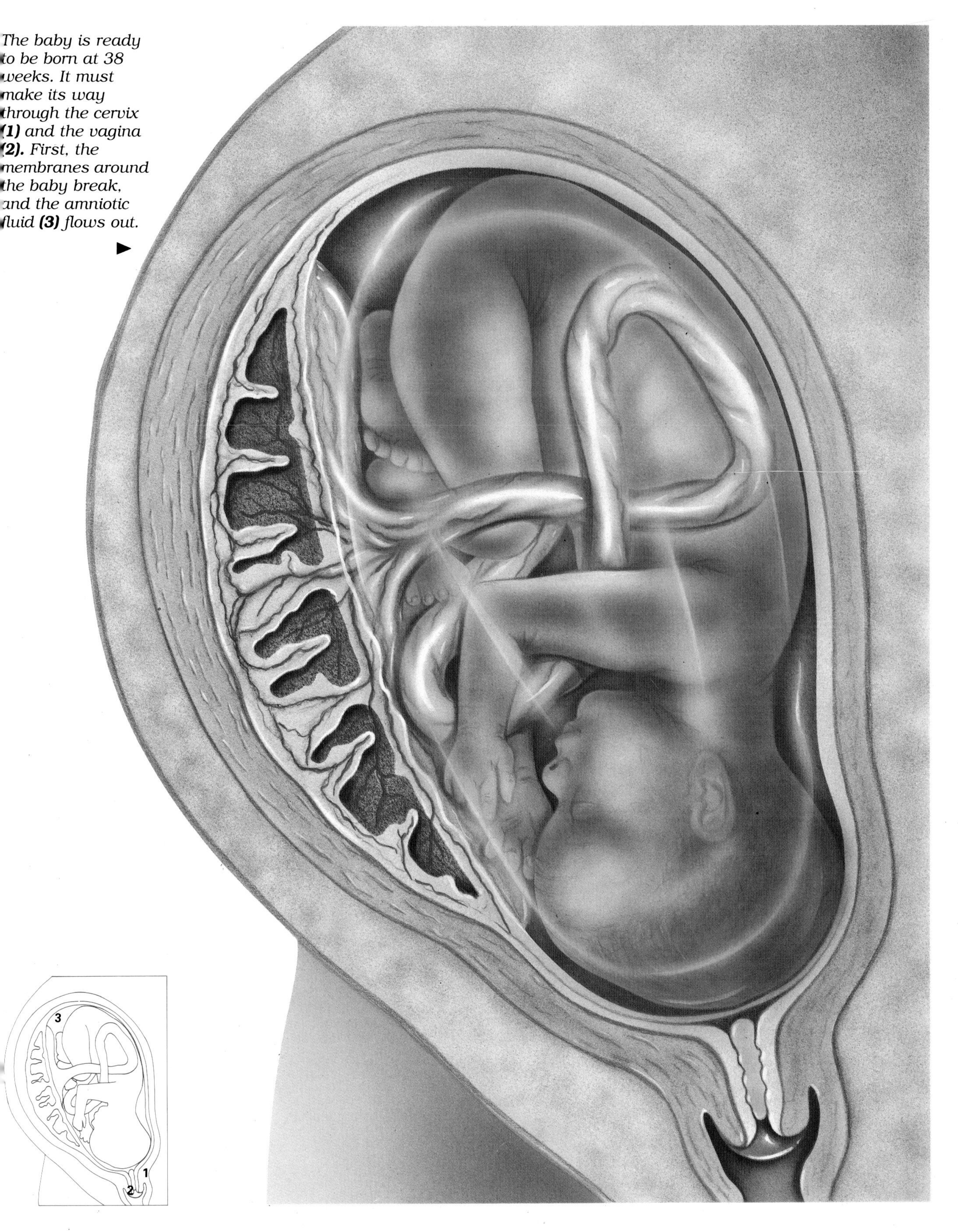

The baby is ready to be born at 38 weeks. It must make its way through the cervix ***(1)*** *and the vagina* ***(2).*** *First, the membranes around the baby break, and the amniotic fluid* ***(3)*** *flows out.*

►

How We Develop

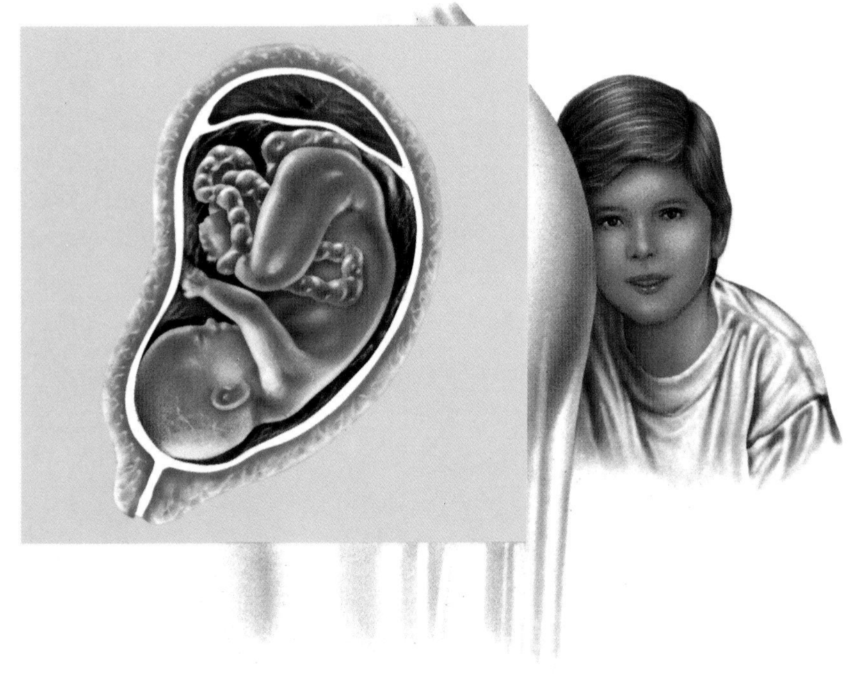

Hearing the Heartbeat

The heart of a fetus beats much more rapidly than the heart of a child or an adult. If a member of your family, or a very close friend, is in the latter months of pregnancy, you may be able to hear the heartbeat for yourself.

To do so, you will need a watch with a second hand. Take the pulse of the pregnant woman by counting how many times her heart beats during one minute. Write this number down.

Now place your ear slightly below the center of her abdomen. You will be able to hear the heart of the fetus. When you count the beats, you will find there are about 250 a minute. Compare the two figures and see how different they are.

Boy or Girl?

Why do some embryos develop into boys and others into girls? The answer lies in the information on the embryo's chromosomes, 23 of which originally came from its father and 23 from its mother. Every woman has two X chromosomes, while every man has one X chromosome and one Y chromosome.

A man can pass either an X chromosome or a Y chromosome to his children; a woman can pass only an X chromosome. Any embryo with a Y chromosome will develop into a male.

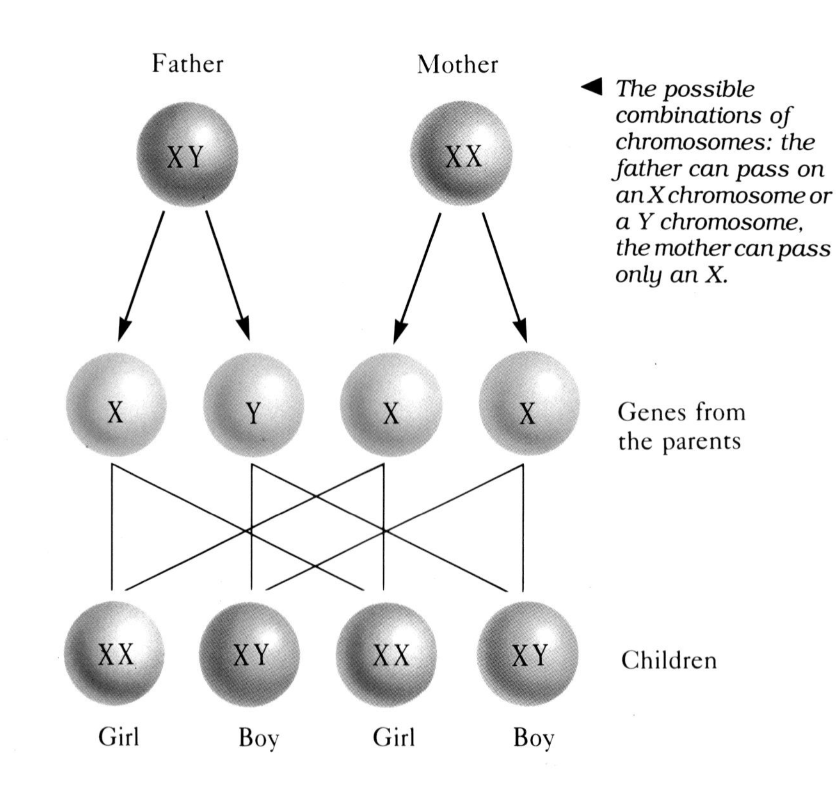

◀ *The possible combinations of chromosomes: the father can pass on an X chromosome or a Y chromosome, the mother can pass only an X.*

The Laws of Heredity

How we develop is ruled by our genes. These are instructions carried on the chromosomes we inherit from our parents, which control the way in which our bodies will take shape.

We carry a complete set of instructions from each of our parents. In many cases, one instruction is stronger, or more dominant, than the other, recessive gene. The gene for dark hair, for instance, is a dominant gene, and that for fair hair is recessive. This means that a person who inherits one gene for dark hair and one for fair hair will have dark hair.

The genes for hair color are passed on from parents to their children. Dark-haired genes are dominant, meaning that a dark-haired father and a fair-haired mother will have a dark-haired child if the father passes on his dark-haired gene. A dark-haired father may have a fair-haired recessive gene, however, and if he passes that gene on, he can produce a fair-haired child. ►

Characteristic	Father's Family			Mother's family		
	Father	Grandpa	Grandma	Mother	Grandpa	Grandma
Color of hair	fair	fair	fair	dark	dark	fair
Wavy or straight	straight	straight	curly	curly	curly	straight
Color of eyes	blue	blue	brown	brown	brown	blue
Shape of nose	pointed	pointed	upturned	button	button	pointed
Shape of lips	broad	thin	broad	thin	broad	thin
Size of ears	small	large	small	large	small	small

◄ *Make a family chart like this one. How do your features compare with your parents' and grandparents' features?*

Glossary

amnion *one of the membranes that surrounds the developing fetus*

blastocyst *an early stage in the development of the embryo*

cervix *the lower end of the uterus*

chorion *the outer membrane enclosing the developing embryo; it produces the placenta*

corpus luteum *a yellow body, formed in the ovary by a follicle which has produced a mature ovum, that secretes the hormone progesterone*

embryo *the developing fertilized egg during the first eight weeks after conception*

endometrium *the lining of the uterus*

epididymis *a coiled tube close to the testicles, in which sperm collect and mature*

estrogen *a sex hormone produced in the ovaries, which stimulates the development of a woman's sex characteristics at puberty and the monthly maturing and release of a fertile egg*

fetus *the developing embryo from eight weeks after fertilization until birth*

labor *the process of giving birth*

meiosis *cell division that results in the production of reproductive cells with half the normal number of chromosomes*

menopause *the period when a woman stops producing a fertile egg each month*

menstruation *the periodic loss through the vagina of blood and cells from the lining of the uterus*

mitosis *the process in which a cell divides into two exact replicas of itself, each containing identical chromosomes*

morula *the cluster of cells formed by the splitting of a fertilized ovum*

ovaries *the reproductive glands in a woman's body that form her ova and secrete the hormones that regulate menstruation and the development of sexual characteristics*

oviducts *also known as the Fallopian tubes, they carry mature eggs from a woman's ovaries to her uterus*

ovulation *the monthly release of a mature ovum from a woman's ovary*

ovum *the female sex cell, or egg*

penis *the male reproductive organ, through which sperm is placed in a woman's body*

placenta *the organ that develops during pregnancy to supply the fetus with oxygen and nourishment from the mother's bloodstream and to carry waste products back to the mother's body for disposal*

prostate gland *found at the base of a man's bladder, it produces the secretion in which his sperm move around*

puberty *the period in the body's development when the reproductive glands and organs start to mature*

scrotum *the pouch of skin containing a man's testicles*

semen *the fluid from a man's testicles and prostate containing his sperm*

sperm *the male reproductive cell*

testicles *the glands in a man's scrotum that produce sperm and male sex hormones*

testosterone *a hormone produced in the testicles that controls a male's sex characteristics*

uterus *the hollow muscular organ in a female's body where the developing embryo grows*

vagina *the passage leading from the cervix to the outside of a woman's body*

zygote *the cell formed when a sperm and ovum unite*

ndex